A fascinating collection of photographs and articles covering some
of the most exciting aspects of
aviation's past and future. A
revealing article on Bomber Command's bad years, 1939-42, by
Sir Robert Saundby, a photo-
graphic essay showing the U.S.
Sixth Fleet on a tour of duty in
the Mediterranean, a look at some
of the lesser-known air museums,
transatlantic mail by catapult,
Greenland's ice-reconnaissance
flights, a short history of the
camouflage used by military aircraft, and air conquest of the
Pacific. Previously unpublished
photographs complement the
histories of the Luftwaffe's three
great fighter squadrons, and in a
story from the Stringbag days an
early aviator tells of learning to fly
in 1918. Flying in India, aviation
news of the past year, and the
latest from space are also
included.

AIRCRAFT
'SEVENTY

AIRCRAFT 'SEVENTY

EDITED BY
John W. R. Taylor

ARCO PUBLISHING COMPANY, INC.
NEW YORK

Contents

page

INTO THE 'SEVENTIES 5
    John W. R. Taylor

NEWS OF THE YEAR 11
    David Mondey

"ARCHIE" 20
    Major W. F. J. Harvey, MBE, DFC and bar, TD

BOMBER COMMAND 1939-42 24
    Air Marshal Sir Robert Saundby, KCB, KBE, MC, DFC, AFC, DL

THE LUFTWAFFE'S "TRADITION SQUADRONS" 30
    André Van Haute

COASTAL COLOURS 39
    Bruce Robertson

WITH THE US SIXTH FLEET 45
    Photographs by Brian M. Service

THEY FOLLOWED THE FLYER 49
    Maurice Allward

STRINGBAG DAYS 55
    E. C. Cheesman

CONQUEST OF THE PACIFIC 59
    Roth Jones

FORGOTTEN FLYING MACHINES 62
    Philip Jarrett

TRANSATLANTIC MAIL BY CATAPULT 66
    Helmut Wasa Rodig

THE HARDEST RECORD OF ALL 70
    Mano Ziegler ... ...

PLEASE DON'T EAT THE PILOT 75
    Sydney Cooper

SOME LESSER-KNOWN AIR MUSEUMS 79
    Leslie Hunt

GREENLAND'S ICE-RECONNAISSANCE FLIGHTS 84
    Carl Christian Brunckhorst

OURNEY ROUND THE MOON 89
    Maurice Allward

[The colour transparency of a Nimrod maritime reconnaissance aircraft used as the basis for the cover design was supplied by Hawker Siddeley Aviation. Blocks appearing between pages 24, 25 and 78, 79 were kindly loaned by Flying Review International and Air BP, the journal of the International Aviation Service of the BP Group]

Frontispiece: First take-off of the British-built Concorde 002 at Filton on April 9th, 1969

Published 1970 by Arco Publishing Company, Inc., 219 Park Avenue South, New York, N. Y. 10003

Into the 'Seventies

by John W. R. Taylor

FEW YEARS in the history of our world have been so momentous as 1969. This is the year in which the man on the Moon ceased to be a mere character in nursery rhymes, when the planets Mars and Venus yielded many of their secrets to space probes, and when travel and communications within Earth's atmosphere prepared to be revolutionised by the supersonic airliner and "jumbo-jet".

Paradoxically, as man demonstrated the extent of his technological progress by investigating other worlds in space, his own world slid backwards. In place of the intercontinental cold war between East and West as the main threat to life and happiness, there was a further reversion to the tribalism of civil war, domestic racial unrest and student riots.

No-one can pretend that the danger of a third world war has passed while any nation can still make a psychological, political and military blunder as great as the invasion of Czechoslovakia by the Soviet Union and its Warsaw Pact allies in 1968. However, the possibility of an all-out nuclear war being started by mistake or miscalculation is now remote. Apart from China (which is unpredictable but in no position to fight the USA or Russia), the nations with atomic weapons know that it would be suicidal to use them, even tactically on a battlefield. The old policy of the "trip-wire" deterrent, with fingers poised over missile firing buttons, ready for instant and crushing retaliation to any seeming threat, has given way to a more realistic plan, at least as far as NATO is concerned.

If any move by the Warsaw Pact countries in Europe should bring them into conflict with NATO forces, the intention is to restrict the clash in both area and types of weapons used. Reaction would be based on checking the attacking forces and striking at their forward bases, in an effort to prevent escalation of the fighting long enough for the statesmen of both sides to consider the consequences of an all-out war and, having done so, to open negotiations for a cease-fire. It is believed that this might take five or six days, and NATO must have strong

A Boeing 747 parked near a 707 emphasises the huge size of the "jumbo-jet"

5

enough non-nuclear forces to contain an attack for that period of time. If the attack could not be held, the use of nuclear weapons would, of course, become imperative, with all the consequences entailed.

The Czechoslovakian crisis left no doubt of how difficult it would be to put this policy into practice. It demonstrated that the Warsaw Pact nations are able to deploy huge forces extremely quickly, by night as well as day. They were able to achieve tactical surprise on this occasion, despite a period of warning of possible invasion. They also made it clear that they will never limit the size of any offensive force to the minimum necessary to achieve its objectives. With immense forces available, they are ready to use a steamroller to crack the proverbial nut.

In itself, the Czech incident did not bring a third world war nearer. The Soviet Union regarded itself as a sort of Socialist parent who needed to chastise its Socialist son for having wrong ideas—and was almost taken aback by the fierceness of the feeling that its action provoked in the west. Nevertheless, the affair provided lessons that NATO has learned well.

The British Government, in particular, has revised its defence planning and military aircraft programmes to a degree that made the 1969 *White Paper* on Defence the most realistic for 14 years. If, for the first time, it sticks to its plans, the aircraft now being ordered and built for the RAF will not only give NATO's new deterrent policy a chance of working but will be equally useful if a future government should reverse decisions such as the planned withdrawal of British forces from east of Suez.

To see why, we need only consider what would happen in a brief East/West clash in Europe. The key weapons would be tactical strike aircraft, able to check the movements of an enemy on the ground and strike back at his air bases. Unfortunately,

any NATO aircraft based on airfields within 350 miles of the frontier would risk the same quick annihilation that came to Arab air forces in the opening minutes of the June 1967 war with Israel.

The only obvious exception would be a VTOL (vertical take-off and landing) aircraft like the Hawker Siddeley Harrier. This unique fighter can be dispersed and camouflaged in fields and clearings well away from concrete runways, and supplied with fuel and weapons by road transport or helicopter. In its initial GR.Mk 1 form, with 19,200lb-thrust Pegasus 101 engine, it has only a short range and small weapon load when required to take-off vertically; but there are few places where load and range could not be increased by using a short forward run. In any case, a more powerful version of the Pegasus is under development, and the Harrier will then combine its VTOL capability and transonic speed with the ability to carry a full $2\frac{1}{2}$-ton load of bombs, rockets and 30mm gun packs over a more normal combat radius.

Present contracts for the RAF total only 90 Harriers, including 13 two-seat T.Mk 2 trainers. Once these have demonstrated their potential in squadron service, it will be surprising if orders do not pour in from all over the world, including the USA.

With three squadrons of Harriers based in Germany and others in reserve in the UK, the RAF will be half-way to meeting the challenging requirements of NATO's new deterrent policy, in terms of survivability if not numerically. To strike back against hostile airfields beyond the Harrier's range, it still needs an aircraft in the class of the cancelled TSR.2 and swing-wing F-111K.

It is to meet this requirement that the United Kingdom has undertaken to develop the MRCA (multi-role combat aircraft) in partnership with West Germany, Italy and—probably, later—the Netherlands and other countries. Announcing this on May

14th, 1969, the Minister of Defence, Denis Healey, said that the RAF will receive about one-third of the initial production series of 1,000 MRCA's, to begin replacing its Vulcans, Buccaneers and Phantoms in the mid-seventies.

What kind of aircraft is able to replace three such differing types, *and* fulfil short-range interception and attack roles in the *Luftwaffe*? At this stage no-one knows. All that has been announced is that the UK and Germany will share design leadership and that the programme will be entrusted to the Panavia Company, formed jointly by BAC, Messerschmitt-Bölkow, Fiat and Fokker, and based in Munich. Even the type of engine to be fitted has not been announced but it would be a tragedy for the European aircraft industry if an American power plant were chosen in preference to a Rolls-Royce design.

The technical success of the Concorde and Jaguar programmes promises well for this international project; on the other hand, the failure of the American F-111B programme emphasises how difficult it is to meet two different combat requirements with one basic design. It would be better to save our money than to allow the need for compromise to produce anything less than what the RAF *really* needs this time.

As a start, the MRCA must have an adequate combat radius to put it beyond the reach of opposing fighter-bombers and still be able to penetrate deeply into hostile territory. It must also be able to elude the most advanced defence system likely to exist by the mid-seventies, when the Warsaw Pact nations will almost certainly have pulse-Doppler radar and anti-aircraft missiles in the class of the British Rapier, to deal with low-flying targets. Equally, the Soviet Air Force should have in service by then a "bomber-destroyer" aircraft, cruising for long periods above 20,000 feet but still able to detect strike aircraft flying at ground level and attack them with "snap-down" missiles.

For any hope of success, the RAF MRCA would probably need to cruise at Mach 0.9 (90 per cent of the speed of sound) at a height of 200 feet, using terrain-following radar for automatic control, active ECM (electronic countermeasures) equipment to confuse the defences, and complex radar for blind attack in all weathers, plus supersonic over-the-target dash capability. We can only wish the Panavia Company luck; it will need it, not only to meet such a specification but to survive the periodical financial crises and cuts, and changes of mind, of several governments.

Transatlantic Race

The *Daily Mail* transatlantic air race, staged in May 1969 to mark the 50th anniversary of the first non-stop crossing by Alcock and Brown in a Vickers Vimy, was billed as "the last great air race". Whether it was either the last or great is debatable; was it even worthwhile?

The answer must be "yes". Whether or not anything was achieved by bowler-hatted businessmen or stewardesses in rickshaws, the fine performances by the RAF and Fleet Air Arm must have helped to convince many youngsters that a career in Britain's flying services still offers excitement and adventure. Even more important, the Harrier demonstrated its capabilities by taking off and landing in the heart of London and New York. No other combat aircraft in the world could do this; it was, therefore, the best possible boost for the Harrier in the export market and the best possible reminder that Britain's aircraft industry can still set the pace in bold, practical design.

Britain's Partnerships

It seems only a few months ago that the press, and others, were predicting the early demise of the British aircraft industry. There was some excuse for this, after cancellation of the TSR.2, HS 681, P.1154, Anglo-French variable-geometry fighter and other important and promising programmes. But events of the past year have shown over and over again that the industry remains one of Britain's greatest assets, technically and economically.

In 1968, its export sales climbed to a record figure of nearly £300million. By March 1969, its total exports since the war had passed £2,500million, of which more than £1,000million came from the sale of aero-engines.

Today, the Anglo-French Concorde supersonic airliner is years ahead of any possible competition from the USA and, if flight development goes well, should earn vast sums in the early seventies. It was a big disappointment when the British Government decided to withdraw from the A.300 European Airbus project in April 1969; but Hawker Siddeley may be retained as sub-contractors for the wing and engine pods of this aircraft, as neither of the surviving partners (Deutsche Airbus and Sud-Aviation) has its experience of designing and building modern transport aircraft.

Few people realise the extent to which the British aerospace industry is contributing to the success of major foreign programmes, For example, Short Brothers & Harland builds outer wings for Phantoms, wings for the Fokker F.28 Fellowship jet-liner and engine pods for the Lockheed TriStar airbus; while Scottish Aviation manufactures whole sections of the fuselage of the Lockheed Hercules transport. This reflects the shortage of US production capacity brought about by the demands of the war in Vietnam; but the only reason for Lockheed's choice of the Rolls-Royce RB.211 turbofan for the TriStar, and for replacement of the original American engine by an Allison-built Rolls-Royce Spey in current

versions of the LTV A-7 Corsair II strike aircraft for the USAF and US Navy, is that the British engines are the best available.

British Lightplane Challenge

Since the war, America's lightplane manufacturers, spearheaded by Beech, Cessna and Piper, have captured the majority of the world market for private, sporting and small business aircraft. Each year, these three companies build some 12,000 single-engined aircraft, of which 25 per cent are exported. Two-thirds of the total are high-wing machines and the Americans are so well established in this market that the big companies in Britain have made no attempt to challenge them.

Down in the Isle of Wight, those two enterprising young businessmen, John Britten and Desmond Norman, decided that it was time to do something about this. Having put their ten-seat twin-engined

City centre jet: Squadron Leader Lecky-Thompson landing his Harrier on a site near the United Nations Building in New York. He checked in at the top of the Empire State Building six hours 11mins 57secs after leaving the Post Office Tower in London, to take two *Daily Mail* air race prizes for the shortest overall time in the East to West direction

The Britten-Norman BN-3 Nymph four-seat lightplane, designed for "knocked-down" export to overseas customers

Soviet Achievements

While Britten-Norman builds 'em small, Russia continues to set new records in giantism. On February 12th, four of the load-carrying records set up by the Mi-6 and Mi-10K were smashed by a new Mil helicopter designated V-12 (Mi-12). Flown by V. P. Koloshenko and a crew of five, it reached a height of 9,678 feet, carrying a payload of 68,410lb. The best way of putting such figures into perspective is to point out that this *payload* of more than 30 tons is equivalent to $1\frac{1}{2}$ times the *complete loaded weight* of the Sikorsky S-64 (CH-54A) Skycrane or S-65 (HH-53B) Sea Stallion, largest helicopters in production outside the Soviet Union.

Among other Russian successes of the past year have been the first flights of the Tu-144 —first supersonic airliner to leave the ground —on December 31st, 1968, and of the shapely Tu-154 tri-jet airliner on October 4th, 1968. In its initial form the Tu-154 will carry up to 164 passengers on medium/long stages, at 528-560mph. A developed version will accommodate 250 passengers.

Up till now, Russia's aircraft industry has achieved no major success in the airliner export field, its best-selling Il-18 notching up no more than 108 sales outside the Soviet Union by the Spring of 1969. The Tu-154 should help to change the picture—not that it really matters when Aeroflot alone needs sufficient aircraft to carry more than 68million passengers annually. With a home market of this size and none of the "balance of payments" problems of a capitalist economy, the Soviet aircraft industry can concentrate on building what Aeroflot and the Soviet armed forces want, without having to think whether the results will pay their way on the export market.

Space

To end this brief comment on a few of the more significant aerospace events of the

BN-2 Islander firmly on the map, with £9million worth of orders and £2million worth delivered, they set out to design and build a four-seat all-metal lightplane. Only 53 days after construction began, the prototype BN-3 Nymph made its first flight on May 17th, 1969, piloted by Norman and with Britten and a flight engineer as passengers.

Like the best-selling cars, the Nymph will be available with a choice of three different engines, rated at 115, 130 or 160hp, so that purchasers can choose the performance category most suitable for the job they want to do—flying instruction, air taxi or executive. Another idea borrowed from the car industry is that Nymphs will be sold in the form of "knocked-down" kits of finished parts, for final assembly under licence in the country where they are sold.

This offers tremendous advantages for everyone concerned. Britten-Norman does not need to set up a costly assembly line. Customers save a lot of money on shipping costs and are presented with an opportunity to start a local aircraft industry, at low cost and with a minimum of skilled labour, in places where no aircraft factory yet exists. Cost of a completed aircraft to the licensee could be cut by anything from a quarter to nearly a half, depending on how many he sold in a year.

First supersonic airliner to fly, on the last day of 1968,
was Russia's Tu-144

past year, what can one add to the millions of words already written and broadcast about the American and Soviet achievements in space? These have been the events of the year, by comparison with which even the first flights of supersonic airliners must seem almost commonplace.

With their Soyuz spacecraft, the Russians have brought nearer the time when scientists will live and work for long periods in Earth orbit, with untold benefit to man's knowledge and wellbeing. With their automatic dockings, they have shown that space stations can be assembled in orbit, without risking the lives of astronauts, clad in fragile space-suits. With TV transmissions from Apollo, America has brought the Moon into our living rooms, showing us scenes that, until a few years ago, none of us could have hoped to see in our lifetime.

What even the superb and nerve-wracking TV commentaries have tended to overlook is that Apollo is not flung towards a stationary object a quarter of a million miles away. As the spacecraft is aimed to skim past the edge of the Moon at a speed of about 5,600 mph, the Moon itself is whirling towards it, on a collision course, at its orbital speed of 2,287mph. A space engineer summed it up as being rather like "running in front of a locomotive close enough to scrape a paint sample off its bumper without getting hurt".

We can only pray that astronauts and cosmonauts will continue to journey into space without getting hurt and that, perhaps, the views they give us of our "good Earth"—which seems to be a beautiful oasis in an awesome, forbidding universe—may inspire us to live here in brotherhood and peace.

News of the Year

Compiled by David Mondey

Mar 2 The first of 58 Lockheed C-5A Galaxies for USAF Military Airlift Command was rolled out at the company's Georgia plant. World's largest aeroplane, the C-5A will lift a 100,000-lb payload over a range of 5,800 miles.

Mar 3 First 24 Mirage III-S's of the Swiss Air Force became operational at Buochs airfield, near Lucerne.

Mar 4 British United Airways ordered five BAC One-Eleven 500s, a "stretched" version tailored specifically to their inclusive tour business, for delivery starting May 1969.

Mar 4 Heaviest and most sophisticated American orbiting geophysical laboratory, OGO-5, was launched by Atlas-Agena rocket from Cape Kennedy.

Mar 6 An Avro C.19 Anson (TX183) was added to the Shuttleworth Collection of historic aircraft at Old Warden Aerodrome.

Mar 8 Third Handley Page Jetstream prototype (G-ATXI), powered by two Turbomeca Astazou XIV engines, flew at Radlett, Hertfordshire.

Mar 9 Mr Peter Scott, well-known ornithologist, writer, painter, broadcaster and glider pilot, became Chairman of the British Gliding Association.

Mar 9 Gen Charles Ailleret, Chief of Staff of the French Armed Forces, was killed in an accident to a Douglas DC-6 at St Denis airport, Reunion Island.

Mar 10 French aviation pioneer, M René Leduc, died at Istres. He is remembered particularly for his work on ramjet-powered aircraft.

Mar 11 Short Skyvan G-ASZI, first to be powered by Garrett AiResearch TPE331-201 engines, landed back at Belfast after successful completion of altitude and high temperature trials.

Mar 13 President of the Board of Trade, Mr Anthony Crosland, announced that British European Airways was to order 26 Hawker Siddeley Trident 3Bs, with an option on ten more.

Mar 17 Six General Dynamics F-111A's of the USAF arrived at Takhli base in Thailand. Operational use over Vietnam resulted in the loss of two before the end of the month.

Mar 18 Minister of State, Mintech, stated that talks had been held between the British government and Canadian, Dutch, German and Italian ministers and officials to discuss British participation in the design and development of a European multi-role combat aircraft (MRCA).

Mar 27 First man in space, Russian cosmonaut Col Yuri Gagarin, was killed when the MiG-15-UTI he was flying crashed at Kirzhatsk, some 40 miles north of Moscow. Killed with him was Col Vladimir Seregin.

Mar 29 Rolls-Royce announced that it had won a £150 million contract to supply RB.211 three-shaft turbofan engines to power the American Lockheed L-1011, a 300-seat tri-jet airbus.

Mar 29 The Royal Danish Air Force ordered 23 Saab Draken 35XDs, comprising 20 single-seat fighter-bombers and three trainers, with deliveries scheduled to begin early in 1970.

April 1 The Royal Air Force celebrated the 50th Anniversary of its formation.

Dassault's swept-wing Mirage F1, designed to replace the delta-wing Mirage III in the French Air Force

April 2 Go-ahead for the Anglo-French helicopter programme was signalled when the British Ambassador in Paris and the French Minister of the Armed Forces signed two Supplements to the Memorandum of Understanding of 1967.

April 4 Marshal of the RAF Sir Dermot Boyle launched an appeal for a £1 million fund to establish a Royal Air Force Museum at Hendon.

April 5 An RAF Hunter flew between the two towers, bascules and upper span of Tower Bridge, London.

April 9 First element of the Netherlands, Belgium and West-German IPG (International Planning Group) automated air-defence system—a programming and training centre—was opened at Glons, Belgium.

April 10 Russian spacecraft Luna 14 was launched into a Moon orbit.

April 13 Martin-Baker Aircraft Company announced that a successful ejection on this date marked the 2,000th life saved by the use of their ejection seats.

April 15 First Boeing 727-200 to be operated outside of the USA—Air France's F-BOJA—started service on the London-Paris route.

April 15 Soviet spacecraft, Cosmos 212 and 213, automatically docked and un-docked in orbit.

April 15 New world distance record for sailplanes set up by Mr James E. Yates in his Schweizer 2-32, with a flight of 681 miles from near Boulevard, California, to Culbertson County airport, near Van Horn, Texas.

April 17 The first prototype Anglo-French Jaguar E-01, a two-seat version for the French Air Force, was rolled out at Breguet's Villacoublay plant in Paris.

April 17 First fully-certificated Hawker Siddeley Trident 2 (G-ANFD) entered service on the London-Milan route.

April 23 Minister of Technology, Mr Anthony Wedgwood Benn, announced in the House of Commons Britain's withdrawal from the ELDO space programme.

April 26 The *Daily Mail* announced an Atlantic Air Race, from the top of the GPO Tower in London to the top of the Empire State Building, New York. To take place between May 4 and 11, 1969, prize money will total £45,000.

April 29 A £100 million-plus Anglo-Libyan package arms deal was announced by British Aircraft Corporation, for the supply, installation, commissioning and support of a complete mobile air defence system, based on Thunderbird and Rapier missiles.

April 29 The first three of the Royal Navy's 28 Spey-engined McDonnell F-4K Phantoms flew into RNAS Yeovilton from the Azores.

April 30 RAF Strike Command was formed officially from the former Fighter and Bomber Commands.

May 1 Hot-air balloon "Bristol Belle" (G-AVTL) made the first ascent at the Hot Air Group's inaugural meeting at the Balloon and Airship Flying Centre at Blackbushe, Hampshire.

May 5 By completing a 3,500-mile flight from Teterboro Airport, New Jersey, to London's Gatwick Airport, a Grumman Gulfstream II became the first executive jet to make a non-stop transatlantic flight.

May 8 A General Dynamics F-111 crashed on a training flight from Nellis AFB, Nevada, following which all USAF F-111As were grounded for investigation.

May 8 Viscount Portal of Hungerford, British Aircraft Corporation's chairman, stated that during the three years ended June 1967, BAC had exported products to a value of over £152 million, and that they had well over £200 million worth of export orders in hand.

May 8 Speaking in the Australian Parliament, Defence Minister, Mr Allen Fairhall, gave assurances that Australia would not take delivery of the 24 F-111s ordered from the US until satisfied of their complete airworthiness.

May 10 Mr Humphrey Wood, commercial director of Hawker Siddeley Aircraft Manchester Division, stated that they had been evaluating market reaction to a proposed 50/70-seat feeder airliner, powered by two Rolls-Royce Trents, with the designation Hawker Siddeley HS 860.

May 13 Prototype of Piper's new commuter airliner, the PA-35 Pocono, made its first flight at Vero Beach, Florida.

May 14 King Olav of Norway opened officially a chain of Decca Navigator stations providing high-accuracy navigational coverage of the whole Norwegian coastline.

May 16   Super VC10 G-ASGK, with 146 passengers aboard, made BOAC's first automatic landing at London Heathrow after a scheduled flight from Chicago and Montreal.

May 16   Carbon-fibre reinforced plastic compressor blades, developed by Rolls-Royce and undergoing trials on the Conway engine, were displayed at a Royal Society meeting.

May 17   Second flight model of ESRO 2, the first European Space Research satellite, was launched at the Western Test Range, California.

May 19   The Royal Swedish Air Force fired its first BAC Bloodhound 2 anti-aircraft missile at the Vidsel test range in Northern Sweden.

May 20   President of the Board of Trade, Mr Anthony Crosland, announced the appointment of Mr Justice Roskill as chairman of a commission to examine the question of London's third airport.

May 22   First Spanish-built Northrop F-5, a two-seat trainer version designated SF-5B, made its first flight at Getafe, near Madrid.

May 23   Chicago O'Hare Airport had a record 2,486 traffic movements during the 24 hours of this day.

May 25   The Grumman EA-6B, a four-seat electronic-countermeasures development of the A-6A Intruder, made its first flight.

June 6   The Society of British Aerospace Companies stated that British aerospace exports for 1968 were 50% up on the same period of 1967 and setting new records.

June 8   Dr Barnes Wallis, 80 years old, responsible for inventions which include geodetic airframe construction, special-purpose bombs and pioneer swing-wing projects, received a Knighthood in the Queen's birthday honours list.

June 10   At RAF Abingdon, a Hawker Hind and a Vickers Gunbus replica were presented for inclusion in the RAF Museum.

June 14   In commemoration of the 50th Anniversary of the formation of the Royal Air Force, Her Majesty Queen Elizabeth II reviewed the Service at RAF Abingdon.

June 17   The first McDonnell Douglas C-9A aeromedical transport for the USAF was rolled out at Long Beach, California.

June 18   Australian destroyer HMAS *Hobart*, operating off Vietnam, was accidentally hit by three Sparrow air-to-air rockets fired by US Phantoms. It has been suggested that they may have homed on to the *Hobart's* radar.

June 18   An unofficial world speed record of 235 mph was set by a Lockheed Cheyenne rigid-rotor helicopter.

June 20   Flt Lt "Tommy" Rose, DFC, who in 1936 set up a London-Cape Town solo record in a Miles Falcon Six, died at the age of 73.

June 20   Inaugural operation by the Rumanian airline Tarom, Bucharest-Frankfurt, was made with a BAC One-Eleven 400 (YR-BCA). Tarom is the first Eastern European airline to operate a new western-type jet aircraft.

June 20   An RAF team won the three-day international rescue helicopter crew competition at Aalborg, Denmark.

June 23   Capt W. E. Johns, creator of "Biggles", world-famous fictional aviation character, died at the age of 75.

June 26   A Phoebus 2A nuclear reactor was tested successfully at the Nuclear Rocket Development Station, Jackass Flats, Nevada. In a 12-min run the reactor generated 200,000lb thrust, marking a milestone in the development of nuclear propulsion.

June 28   Six Avro Anson C.19s of Southern Communications Squadron at Bovingdon flew past in formation to mark the "Annie's" retirement after 32 years service with the RAF.

June 30   The Lockheed C-5A Galaxy made its maiden flight from Dobbins AFB, Georgia, piloted by Leo Sullivan, Lockheed-Georgia's chief test pilot.

June 30   A Seaboard World Airlines DC-8, on charter to USAF Military Airlift Command, carrying 214 US troops and Government employees, was forced by Soviet fighter aircraft to land on the Russian island of Iturup.

July 11   A Supplementary White Paper on Defence stated that 26 Hawker Siddeley Buccaneer S.2s were to be ordered for the RAF.

July 13   First production model of 210 General Dynamics FB-111s on order for USAF Strategic Air Command flew at Fort Worth, Texas.

July 15 Moscow-New York service, via Montreal, inaugurated by Aeroflot with Ilyushin Il-62s. A Boeing 707 of Pan American started a reciprocal service from Kennedy International Airport, via Copenhagen.

July 17 Representatives of Belgium, Canada, the Federal German Republic, Italy, the Netherlands and United Kingdom, met in Bonn to decide upon forms of co-operation for an Advanced Combat Aircraft (ACA) for common military use.

July 19 A new Australian-British contract was signed for continuation of joint projects at Woomera.

July 30 HRH Prince Charles began a series of flying aptitude tests in a Chipmunk at RAF Tangmere, with Sqn Ldr Phillip Pinney—a CFS instructor—as pilot.

Aug 4 Israeli Air Force Skyhawk fighter-bombers were in action against Jordanian guerrilla bases 15 miles from Amman.

Aug 5 A new STOL strip was opened for operations at New York's La Guardia airport.

Aug 10 Dr George E. Mueller, Associate Administrator, Manned Space Flight, NASA, spoke about new thinking by NASA on the design of re-usable launch systems at a meeting of the British Interplanetary Society.

Aug 16 British European Airways signed an £80 million contract with Hawker Siddeley Aviation for the supply of 26 Trident 3B's.

Aug 18 Concorde prototype 001 started taxying trials.

Aug 18 Russia's prototype Tupolev Tu-154 trijet was rolled out. Bigger than either the HS Trident or Boeing 727, it has "stretch" potential for up to 200 seats.

Aug 20 Czechoslovakia was occupied by troops of the USSR and other Warsaw Pact countries. Ruzyne Airport, Prague, was closed to international services.

Aug 23 First McDonnell Douglas F-4M Phantom for the RAF was delivered to No. 228 OCU at RAF Coningsby, Lincolnshire.

Aug 24 King's Cup Air Race was won by Ron Hayter, flying a pre-war de Havilland Hornet Moth, at an average speed of 121mph.

Aug 28 M Robert Morane, a pioneer pilot and founder of the French Morane-Saulnier aircraft company, died in Paris at the age of 82.

Aug 29 British European Airways took delivery of the first of its fleet of 18 BAC One-Eleven 500s at Hurn, a month ahead of schedule.

Aug 31 The Rolls-Royce RB.211 turbofan engine completed successfully a first test run at Derby.

Sept 4 The first F-111C for the RAAF was handed over to the Australian Minister of Defence at Fort Worth, Texas.

Sept 5 The 3,000th McDonnell Douglas Phantom was delivered to the US Navy.

Sept 8 The prototype Anglo-French two-seat trainer/tactical support Jaguar made a successful first flight from the Centre d'Essais en Vol at Istres, piloted by Breguet's chief test pilot, M Bernard Witt.

Sept 8-9 Max Conrad, the 65-year-old pilot, claimed a closed-circuit distance record of 4,968 miles.

Sept 11 The Dassault MD-320 Hirondelle made its first flight at Bordeaux Merignac, piloted by M Hervé Le prince Ringuet.

Sept 12 The second prototype BAC/Sud-Aviation Concorde, 002 (G-BSST), was rolled out at Filton.

Sept 13 Dowty Rotol announced an order worth £2.9 million for the supply of 150 nose-wheel gears for the McDonnell Douglas DC-10.

Sept 15 Russia launched her Zond 5, an unmanned lunar probe.

Sept 18 Avions Marcel Dassault and Ling-Temco-Vought Aerospace Corporation announced an agreement to exchange technical information on variable-geometry aircraft.

Sept 18 The McDonnell Douglas DC-9-20 made its first flight from Long Beach, California, after a take-off run of only 2,400 ft.

Sept 21 Russia's Zond 5 splashed-down in the Indian Ocean after completing successfully a circumlunar flight.

Sept 24 BEA and Air France signed an agreement to pool services on German domestic routes as from 1 April 1969.

Sept 24 First emergency ejection from a German Air Force F-104G Starfighter equipped with a Martin-Baker GQ7A seat.

Sept 26 The first LTV A-7D Corsair 2, powered by a Rolls-Royce/Allison TF41-A-1 (Spey) of 14,500lb thrust, made its first flight.

Sept 30 The prototype Boeing 747 was rolled out at the company's new factory at Everett, near Seattle.

Above: Artist's impression of the Hawker Siddeley Trident 3B, with RB.162 turbojet above central engine to boost take-off performance

Left: The distinctive shape of the Lockheed C-5A Galaxy, largest aircraft ever flown

Right: Lockheed's L-1011 TriStar is a 244/345-seat airbus, for service in 1971

Sept 30	Aircraft designer and engineer, C. C. Walker, one of the five founder members of the de Havilland Aircraft Company, died at the age of 91.	
Sept 30	First police force in the world to make regular use of helicopters, New York City Police celebrated the 20th anniversary of their inauguration into daily use.	
Oct 1	First BEA Vanguard (G-APEM) for conversion to all-cargo Merchantman standard was delivered to Aviation Traders (Engineering) at Southend.	
Oct 4	BUA (Holdings), parent company of the BUA group, announced the formation of a new airline, British United Island Airways, comprised principally of BUA (CI), BU(Manx)A and Morton Air Services.	
Oct 4	The Russian Tu-154 trijet made its first flight piloted by Nikolai Goryainov.	
Oct 9	The United States began negotiations for the sale of 50 McDonnell F-4 Phantoms to Israel. This followed French refusal to supply Israel with 50 Mirages ordered previously.	
Oct 9	First round-the-world flight by a Short Belfast, of 53 Squadron Air Support Command, was completed in a total flight time of 88 hours.	

Oct 11   Launch of America's first manned Apollo spacecraft, carrying astronauts Walter Schirra, Walter Cunningham and Don Eisele on an 11-day Earth-orbital flight.

Oct 14   Minister of State, Mintech, stated that the dollar loss to Britain resulting from the cancelled F-111 order was likely to be about £25 million.

Oct 16   It was announced that the prototype Anglo-French Jaguar had flown supersonically and logged 14 flights since first flight on 8 September.

Oct 21   First General Electric CF6 turbofan engine, the power plant intended for the McDonnell Douglas DC-10, started tests two weeks ahead of schedule.

Oct 22   Apollo 7 splashed down in the Atlantic after a highly-successful 163-orbit 260-hour flight.

Oct 23   A Mirage IV bomber, which had been acting as a Concorde flight simulator, crashed in France. Both members of the crew made successful ejections in their Martin-Baker seats.

Oct 24   Last flight of the X-15 research programme was made by NASA test pilot William H. Dana.

Oct 26   Russia launched her Soyuz 3 spacecraft into Earth orbit, carrying cosmonaut Col

Georgy Beregovoi. During the first orbit he accomplished a rendezvous with the unmanned Soyuz 2, which had been launched the previous day.

Nov 1   First public hearing of the Roskill Commission on the Third London Airport started in London.

Nov 4   The Czechoslovakian prototype Aero L-39 advanced jet trainer flew for the first time.

Nov 6   British Eagle, one of Britain's leading independent operators for more than 20 years, ceased operations.

Nov 6   Mr Denis Healey, British Minister of Defence, speaking at the British Atlantic Committee dinner, said that " . . . negotiations in which Britain is now engaged with Germany, Italy, Holland and Belgium on the possible development of a new advanced combat aircraft (ACA) for the later 70s have major implications not only for strategy but for European industry and technology as well."

Nov 6   The new Heathrow Terminal No. 1 was opened for limited operations by Mr Peter Masefield, chairman of the British Airports Authority.

Nov 7   Mr Henry Ziegler, President of Sud-Aviation, said that the Concorde 001 prototype would make its first flight in January, 1969.

Nov 12   Loftleidir Icelandic Airlines began operating its low-fare North Atlantic services via Iceland, through Gatwick, giving this latter airport its first scheduled transatlantic services.

Nov 12   NASA announced that the Apollo 8 mission, scheduled for 21 December, would be a manned Moon-orbiting flight.

Nov 13   First successful flight of the rocket-powered Northrop HL-10 lifting-body research vehicle made at Edwards AFB, California, piloted by NASA research pilot, John A. Manke.

Nov 14   NASA announced plans to launch two Mariner-type spacecraft to orbit Mars in 1971.

Nov 14   British Minister of Defence, Mr Denis Healey, stated that an additional 20 HSA Harriers were being ordered for the RAF, to permit NATO to assign a further squadron to be based in Germany.

Nov 16   Russia launched Proton 4, world's largest unmanned spacecraft, into Earth orbit.

Nov 19   Only known surviving example of a Hawker Typhoon fighter was handed over at RAF Shawbury for inclusion in the RAF Museum.

Nov 21   First prototype HP Jetstream 3M made its first flight from Radlett, flown by Chief test pilot John Allam. This is the Garrett AiResearch TPE331-powered military version, of which 11 have been ordered for the USAF.

Nov 27   Minister of State, Mintech, stated in Parliament that a feasibility study on a multi-rôle combat aircraft (MRCA) is being carried out by the British aircraft industry following collaboration discussions with Germany, Italy, and the Netherlands.

Nov 28   West Germany's decision to buy 88 McDonnell RF-4E Phantoms was approved by the finance committee of the Bundestag. They also approved purchase of an additional 50 F-104G Starfighters.

Nov 28   Transglobe Airways of Gatwick, another independent airline, announced its closure with commitments running into "several hundred thousand pounds."

Nov 30   Miss Sheila Scott was presented with the Royal Aero Club's Britannia Trophy for the most meritorious performance in the air during the previous year.

Dec 3   First HP Jetstream for the home market was flown to Oxford Airport, Kidlington, for finishing to the requirements of its new owners, Clarke Chapman and Co of Gateshead.

Dec 5   HEOS (highly eccentric orbiting satellite) was launched successfully from Cape Kennedy by a Delta rocket. HEOS is ESRO's most advanced satellite and Europe's first deep-space probe.

Dec 8   The second Bell lunar landing research vehicle crashed and was destroyed at Ellington AFB. NASA chief test pilot, J. Algranti, made a successful ejection.

Dec 9   Minister of State at the Board of Trade Mr William Rodgers, opened the new cargo tunnel connecting the central area of Heathrow with the cargo terminal area.

Dec 11   Sud-Aviation and Hawker Siddeley announced preliminary details of a smaller, less costly to develop version of the Airbus, designated A-300B, made possible by introduction of the Rolls-Royce RB.211 turbofan engine.

Dec 12   British Minister of Technology, Mr Anthony Wedgwood Benn, commenting on the proposal for the A-300B, said that the situation would need careful consideration, his tone suggesting that Britain might withdraw from such a venture.

Dec 18   First Intelsat 3 communications satellite launched successfully by Delta rocket from Kennedy Space Center.

Dec 18   Mr Denis Healey, Minister of Defence, stated that Britain might go-it alone on an MRCA if it proved impossible to design an aircraft able to meet the requirements of all the partners in the project.

Dec 18   Air Chief Marshal Sir James Robb, GCB, KBE, DSO, DFC, AFC, died at the age of 73. He was Deputy Chief of Staff (Air) to Gen Eisenhower in 1944-45.

Dec 21   American spacecraft Apollo 8 was launched to the Moon by Saturn V rocket from Cape Kennedy—man's first-ever deep space flight—with astronauts Col Frank Borman, Capt James Lovell and Maj William Anders on board.

Dec 24   Apollo 8 spacecraft, by then comprising the command module and service module, was injected into an orbit around the Moon, at an altitude of 69 miles above the lunar surface.

Dec 25   Apollo 8 spacecraft, after completion of 10 orbits around the Moon, was injected into an Earth-return path.

Dec 26   Arab gunmen made an attack on an El Al Boeing 707 at Athens International Airport; one person was killed.

Dec 27   Apollo 8 spacecraft, with service module jettisoned, touched down safely in the Pacific Ocean, a mere 5,000 yards from the recovery carrier USS *Yorktown*.

Dec 28   Israeli forces made a helicopter-borne commando attack on Beirut International Airport, in reprisal for the 707 attack at Athens. It resulted in the destruction of 13 aircraft of three Lebanese airlines, which together with damage to airport installations, involved a loss of some £22 million.

Dec 31   The Russian Tupolev Tu-144 SST prototype made its first flight, commanded by test pilot Eduard Elyan, with Mikhail Kozlov as assistant test pilot and two engineers on board.

Jan 3   First Handley Page Jetstream for the home market was delivered to CSE (Aircraft Services) Ltd at Kidlington, for equipping to customer's requirements.

Jan 5   An unmanned planetary probe, Venus 5, was launched from the Soviet Union. It was intended to make a soft-landing on Venus in mid-May.

Jan 7   The 1,000th Northrop T-38 Talon supersonic trainer for the USAF entered service.

Jan 9   Neil A. Armstrong, Lt Col Michael Collins and Col Edwin E. Aldrin were named as the lunar landing crew for America's Apollo 11.

Jan 10   A second Russian spacecraft intended to soft-land on Venus in mid-May, Venus 6, was launched from Baikonour.

Jan 14   BAC announced receipt of a contract from Mintech, on behalf of the Ministry of Defence, for the supply of more than 100 Jet Provost Mk 5 basic trainers for the RAF.

Jan 14   Russia launched a manned spacecraft, Soyuz 4, into Earth orbit carrying Lt Col Vladimir A. Shatalov.

Jan 14   The Royal Navy's first F-4K Phantom training unit, No 767 Squadron, was commissioned at Yeovilton, Somerset.

Jan 14   The USS *Enterprise*, the world's largest warship and only operational nuclear-powered aircraft carrier, was damaged severely by explosion and fire when engaged in exercises off Pearl Harbor. Some 15 crew members were killed, 100 injured and 15 aircraft destroyed.

Jan 15   Soviet spacecraft Soyuz 5 was launched from Baikonour carrying three cosmonauts: Lt Cdr Boris Volynov, Lt Col Yeugeny Khrunov and Alexei Yeliseyev.

Jan 16   Soyuz 4 and 5 were docked and mechanically coupled in Earth orbit. Cosmonauts Khrunov and Yeliseyev transferred to Soyuz 4, using portable life-support systems.

Jan 17   Soyuz 4, containing 3 cosmonauts, was brought in to a successful landing at Karaganda, some 375 miles from Baikonour.

Jan 18   Soyuz 5, with cosmonaut Volynov aboard, landed safely near Baikonour.

Jan 20   The Minister of Defence, Mr Denis Healey, told the House of Commons that: "The feasibility study for the Anglo-German VG aircraft would be completed by the end of the month and

Jan 22 would be considered by the participating countries during February and March".

Jan 22 OSO 5, NASA's fifth orbiting solar observatory, was launched by Delta rocket from Cape Kennedy.

Jan 29 In answer to questions in Parliament on the A-300B and BAC Three-Eleven projects, Mr Anthony Wedgwood Benn, Minister of Technology, stated that: ". . . .there can be no advance commitment to government support for either aircraft".

Jan 30 Mintech announced that a Space Division was to be set up, to integrate administrative and technical branches dealing with space affairs.

Jan 30 Canadian-built ISIS-1 satellite launched by Delta rocket from the NASA Western Test Centre, California, as part of a joint Canadian-American research programme.

Jan 31 Explorer 1, first satellite launched by America, completed its eleventh year in orbit.

Feb 2 A £20 million contract—negotiated by BAC—for the installation, commissioning and initial maintenance of Libya's Thunderbird/Rapier air defence system, including training of personnel, was signed in Tripoli.

Feb 4 A $40million contract was awarded to Grumman Aircraft Corporation by the United States Navy for development of the F-14A variable-geometry fighter.

Feb 6 The three development Lockheed C-5A Galaxies had accumulated a total of 47 flight hours at this date.

Feb 9 The Boeing 747 prototype made a successful maiden flight from Paine Field near Everett, Seattle, piloted by the company's 747 project pilot, Mr Jack Waddell.

Feb 9 Tacsat 9, claimed to be the world's largest communications satellite—with a capacity of 10,000 two-way telephone conversations—was launched successfully from Cape Kennedy by Titan 3C rocket.

Feb 11 The second prototype Anglo/French Jaguar made its first flight from the French test centre at Istres, piloted by Breguet's chief test pilot, M Bernard Witt.

Feb 12 It was announced that a Russian helicopter designated V-12, piloted by V. P. Koloshenko, had carried a load of 68,410lb—over 30 tons—to a height of 9,678ft.

Feb 15 The Chairman of America's Eastern Airlines said that Eastern had given airframe and engine manufacturers their specification for a 100/150-seat 500-mile range STOL aircraft, and were prepared to place orders immediately such an aircraft could be delivered at an economic price.

Feb 17 Flight Shuttle Inc of America inaugurated an on-demand helicopter air taxi service between Manhattan and Kennedy International or Newark Airports with four-seat Bell JetRangers.

Feb 25 Mariner 6 was launched by Atlas-Centaur rocket from Cape Kennedy. It is due to reach the vicinity of Mars on July 31st, 1969, and forms part of a NASA programme to increase substantially our knowledge of Earth's sister planet.

Feb 26 Mr W. G. Carter, CBE, designer of the Gloster/Whittle E.28/39, which led to the highly-successful Gloster Meteor, died at the age of almost 80 years.

Feb 27 The Boeing 747 prototype, after completion of nearly 12½ hours flight testing at Paine Field, Everett, was flown to Boeing Field, Seattle, for the next stage of the development programme. To date the 747 had been flown at weights up to 520,000lb—over 230 tons.

Mar 2 The first prototype BAC/Sud-Aviation Concorde 001 made a completely successful maiden flight at Toulouse, piloted by M André Turcat, Sud's chief test pilot.

Mar 3 Apollo 9 spacecraft, containing all the elements for a lunar landing, was launched from Cape Kennedy on a ten-day flight with a crew comprising Col James McDivitt, Col David Scott and Mr Russell Schweickart, on what was regarded as the most advanced manned flight to date.

Mar 3 The Roskill Commission announced its short list of sites for a third London airport. These were Cublington (formerly RAF Wing), Thurleigh (RAE Bedford), Nuthampstead, Essex and Foulness, offshore Essex.

Mar 7 The Swiss/Italian prototype AS 202 Bravo aerobatic trainer, built by the Swiss company of Flug und Fahrzeugwerke AG and Siai–Marchetti of Italy, made its first flight at Altenrhein piloted by Manfred Brennwald.

Mar 9 Mr Ernest Brooks, designer of the Brookland Mosquito Gyroplane, was killed when he crashed in one of his aircraft at Tees-side Airport, Co Durham.

Mar 13 The Apollo 9 command module splasheddown in the Atlantic after a ten-day flight in which all the scheduled exercises had been completed successfully.

Mar 14 The British-built BAC/Sud-Aviation Concorde 002 commenced taxiing trials at Filton.

Mar 14 The Lockheed-Georgia XV-4B Hummingbird 2 jet-lift research aircraft crashed at Marietta, Georgia, shortly after take-off. The Lockheed test pilot, who ejected at 6,000ft, suffered only minor injuries.

Mar 18 Air Support Command was involved in an operation to carry troops and equipment to the Caribbean island of Anguilla, a total of 16 aircraft being used.

Mar 19 The United States Secretary of Defense stated that the US Marine Corps plans to order 12 Hawker Siddeley Harriers.

Mar 19 Nordair of Montreal inaugurated the first scheduled jet service inside the Arctic Circle, with a weekly return flight on a 2,300-mile route from Montreal to Resolution Bay, Cornwallis Island, using Boeing 737-200s.

Mar 20 The second development Dassault Mirage F1 attained a speed of Mach 1.15 on its first supersonic flight from the Istres Test Centre.

Mar 24 The Dassault Mirage F1 attained a speed of Mach 2.03 on a test flight from Istres.

Mar 28 BAC stated that: "Four European aircraft companies, BAC, Fiat, Fokker and Messerschmitt-Bölkow, have announced the formation of Panavia Aircraft GmbH. The objects of this new company are the management and performance of contracts for the study, development, production and marketing of the multi-role combat aircraft system . . . . "

"Archie"

**by
Major W. F. J.
Harvey**
MBE, DFC and bar, TD

IN THE DAYS before most of us began to take everything so seriously, there was an eccentric English form of humour which derided peculiar or dangerous things, or situations, by labelling them with ridiculously inappropriate names. That is why the Royal Flying Corps referred to German anti-aircraft (AA) guns and their products as "Archie", taken it is believed from the key words, "Archibald, certainly not", of a music hall song by that eminent comedian George Robey. The Germans, being more serious-minded, used the abbreviation *'Flak'* —short and sharp, like their AA.

For some undiscoverable reason, Germany was earliest in the AA field, with the French next and Britain (as usual in preparations for war) last. In *The New Art of Flying* by W. Kaempffert, published in 1910, there are photographs of three heavy AA (HAA) guns made by Krupps, with performance details as follows:—

6·5-cm calibre—shell 8lb 13oz—max height range 18,700ft.

7·5-cm calibre—shell 12lb 2oz—max height range 18,950ft. (lorry borne).

10·5-cm calibre—shell 40lb—max height range 34,000ft.

The height range given for the 10·5-cm shell was a modern instance of 'drawing the long bow', and that to breaking point: no AA gun ever had such a range.

In 1914 Germany went to war with 36 HAA guns. Britain of course had none; although a one-pounder pom-pom was, for some obscure reason, soon mounted on the roof of the Foreign Office in London. By the end of the war Germany possessed 2,576 AA guns, and there were 364 with the British Army in France. At this point it should be made clear that only these heavy guns and their bursting shells were "Archie", as distinct from all other forms of anti-aircraft frightfulness.

During flying training in England, little or nothing was told us about Archie, or our future acquaintance with him; and one was much too busy coping with ordinary flying hazards to ask. Most people had been caught in raids on London or elsewhere, but the guns were thought of merely as noisy and useless vulgarians whose falling waste products were sharp and spiteful. So, during one's first flight over the lines after arrival in France, the sight and sound of a nearby group of those greasy-looking black balls of

---

The sketch of a B.E.2C being chased by *'Flak'*, at the top of this page, is from a collection of paintings by Sir Robert Saundby, published in book form under the title *Flying Colours* in 1918

smoke, each with a core of infernal redness, was something of a shock. The present writer's first offensive patrol at 18,000 ft seemed punctuated by sharp 'cracks' or 'crumps' (i.e. very near, or not so near) above the noise of the open exhausts, and I landed with some unpleasant-looking tears in the fabric, which made me think.

Some flying officers disliked Archie more than Huns, but in course of time one learned how to make things more difficult for him. After escaping his first brackets, I found myself developing a kind of love/hate feeling, almost proud of his skill and the excellence of his equipment. I visualised the battery commanders of those units nearest the lines— in particular he who exercised his craft from sites round the Fabrique Metallurgique at La Bassée—as earnest professors of mathematics, with heads a little too large for their bodies, pebble-spectacled and dedicated to the solution of abstruse problems, Euclid in one hand and a time-fuze and correction table in the other.

In the latter half of the 1914/18 War in France, Archie's deployment was fairly rigid, and he rarely opened up from unexpected places. His first line of defence was a mile or so behind the trench line, the batteries' arcs of fire usually overlapping. Best remembered were those opposite Ypres and Armentiéres at La Bassée, and at Vitry opposite Arras, guarding the vital road junctions at Bapaume and Peronne (when in German occupation), and at St Quentin.

Another concentration was around their balloon line, whose defence was shared by the first-line AA. The flying height of each balloon was, naturally, known to an inch to the enemy gunners, which made balloon strafing considerably more perilous than fighting Huns. Still further back, the network of airfields such as those around Douai, which was also a German Army HQ and rail junction, was heavily defended. And, finally, Archie was to be encountered over

vital rail junctions 10 to 30 miles back, including Courtrai, Lille, Somain, Valenciennes, Cambrai and Busigny.

During battles of some movement in the latter part of the war, Archie learnt to be mobile and was often found to be well up behind his front-line troops. On the whole he always knew the form and what to expect; and so did we when we became experienced, like old cock pheasants at the start of the shooting season.

He knew his regular customers and what they intended doing, and had a very good idea of where they would do it. I am almost certain that he knew individual flight commanders, their peculiarities, and their probable reactions to his assaults. Those on our front must certainly have known the three flight commanders of my squadron after we acquired the first radio telephones used in war—equipment that the German High Command was most anxious to recover from a shot-down machine.

However light-heartedly he was regarded— in a wary sort of way—Archie was not to be despised and certainly could not be ignored. He got his first hard training during the four months of the Somme battle of 1916 when, for the only time during the war, we had mastery of the air. In only one week, more than 2,000 RFC machines crossed over into his orbit. From then on his skill increased until, by the end of the war, he had claimed over 1,500 aircraft. This figure must be accepted as accurate because the HAA layout was such that almost all came down in enemy-occupied territory. The number shot down by ground troops (other than AA units) is not included or known.

I have an abiding memory of his expertise, noted at the time, when I encountered and went for one German balloon flying at the unusual height of nearly 7,000ft. The observers in its basket were obviously reporting by telephone the progress of our great attack of August 21st, 1918, which was the beginning of the end. On nearing the gasbag I found myself ringed by a wall of black bursts, and, circling, saw it begin to contract towards the centre. Only my favourite escape route from trouble, in the shape of a vertical sideslip, saved me.

In course of time Archie developed several sidelines. He became good at 'pointing' our aircraft for the offensive or defensive benefit of his own, the black bursts of his shells being more visible to the enemy pilots than the semi-opacity of distant aeroplanes. After the pointer bursts, the next salvo consisted of bursts to indicate the number of our machines. Some of his guns were also known to have acted in an anti-tank role in moments of crisis; as, occasionally, German field guns assumed the role of Archie against low flyers. For instance, when the Portuguese Division holding a small sector of the line expected a sudden attack one evening and called for help, four of us who were just starting dinner in the Mess took off in the dusk and shot up the roads behind the German front, from under 1,000ft. The spread of flashes from gunfire showed clearly that field guns had joined in, aiming at our flaming exhausts— quite ably, too, for in a letter that night I remarked "My word, the Huns did get annoyed. I have never seen so much Archie. The bursts quite lit up my instruments".

Archie had occasional eccentricities. Once some of his shell bursts were coloured red or green, seriously disturbing the mental balance of one of our pilots at whom they were directed. At another time, there were rumours of 'Pink Austrian Archie' which apparently were of deep significance to our General Staff, for there was a rumoured promise of the award of an MC to any pilot who produced conclusive evidence (though how, was not said). One day I did actually see two pink bursts, but no MC followed.

One day in 1916 Archie produced an offspring, an ill-favoured creature first encountered by my squadron on 22nd September. It consisted of ten or twelve balls of burning phosphorus strung together on a wire which came up almost vertically, wriggling as it came, and intended to wrap itself round any aircraft in its way. They were quite harmless if seen in time—rather like coming across an adder in the undergrowth—but were used until the end of the war. I need hardly add that they were immediately dubbed "flaming onions".

Archie had another ungentlemanly relative whom everyone agreed to dislike. There was no sporting element in machine-gun fire from the ground against our low-flying aircraft, particularly after the enemy learned about deflection shooting. Early in the war all troops of both sides fired on all aeroplanes, friend or foe; but only the bad shots had any chance of hitting, and that was rare. But when individuals of the RFC took to ground strafing in 1916, the German Staff sat up and set about training their troops in light AA (LAA) work. So seriously did they take this new form of warfare that a captured Army Order stated: "An award of 200 marks (about £9) is being given to private soldiers who shoot down British aircraft by fire from the ground. (Not given to AA units)".

The effectiveness of these measures was shown during the month following March 21st, 1918, when two great German advances nearly ended in their victory. Of the 1,232 front-line RFC machines on charge to squadrons at the beginning, 1,032 were written off in four weeks; those who were there would agree that the majority were lost to rifle and machine-gun fire during our intensive ground attacks. The writer's Bristol, one of the 200 which survived, was described in a letter written at the end of this period as "my poor old machine . . . . all patches where nasty Huns have been firing from the ground", and this was normal.

During any low attack on the front line or

Another sketch from *Flying Colours*, showing French Caudron twin-engined reconnaissance aircraft under attack by Albatros fighters amid "stale" anti-aircraft bursts

other heavily defended position with trained LAA personnel, at least one enemy machine-gun could put up close-range, non-deflection fire. The aircraft was then a 'sitter' no matter how violently it performed evasive manoeuvres, because we were flying at comparatively low speed in unarmoured machines in face of swift, short-range bullets. But there was little evidence to show that an AA weapon had been designed to fill the gap between 2/3,000ft, below which our contact and low reconnaissance aircraft operated and where Archie was at a technical disadvantage, and 5/6,000ft where our artillery observation machines flew their figure-of-eight courses over the lines while correcting the fall of shot. But I have one photograph of what appears to be a heavy machine-gun, lorry mounted; and one of my combat reports commented: "On the way back to the line, I saw machine-gun tracer (brown smoke) from the ground pass vertically by me at about 6,000ft". This was far beyond the range of the standard gun.

There were, of course, other kinds of AA, in the strict sense of that word. There was the apparently innocent observation balloon, its basket occupied by 500 lb of ammonal, which was exploded electrically from the ground as the attacking machine drew near. One of these booby-traps was touched off by my rear gunner's long-range shots when he was not really trying. Then there were the barrage balloons surrounding Rhineland towns whose cables discouraged low bombing, particularly at night. The crew of one F.E.2b had a wing bitten into and were saved by the pilot keeping his throttle slightly open, so that the machine spun round and round down the cable until it sat on the ground with a bump.

*　　　*　　　*

Dear Archie! In one's sere and yellow age, when all is forgiven but not forgotten, it is heartwarming to reflect on the tens of thousands of pounds which I, like others, must have cost the enemy for the, literally, tens of thousands of shells addressed to me personally during 150-odd offensive patrols and escorts. I picture all the worn gun barrels, the pay and allowances of those professors of maths, and their myrmidons; and the miles of sausage they must have eaten to keep up their strength. In particular, I remember my old enemy at La Bassée who one day lost his temper and so changed from a nebulous menace to something human and three-dimensional. I had led my formation into a trap where we picked off two of the bait, and then escaped a few seconds before the arrival of the hunters from the sun. Sliding rapidly earthwards ahead of them, we shot up Archie's pitch in passing and bolted across the lines, chased by his thunderballs. When just out of his range, we flew up and down watching him vent his feelings on the empty air.

Sometime in 1918 I wrote home: "The Hun AA gunners are quite old friends now. I am thinking about dropping a note asking for their photographs one day." Come to think of it, how appropriate if my professor's name really was Archibald, or its German equivalent.

Bomber Command 1939-42

by Air Marshal Sir Robert Saundby, KCB, KBE, MC, DFC, AFC, DL

ON DECEMBER 14th, 1939, twelve Wellington bombers, attacking enemy warships in daylight in Schillig Roads, were intercepted by German Messerschmitt Bf 109s, with the result that five were shot down and one crashed when it reached its home base. A similar attack on December 18th suffered an even worse fate. Twenty-two Wellingtons were heavily engaged by German fighters, ten were shot down and three more crashed on return. There were several reasons for this failure.

At the outbreak of the war in September 1939 the British and French Governments announced that it was their policy to limit attacks from the air to strictly military targets, where no errors of aim could possibly cause civilian casualties. In practice, this restricted Bomber Command to attacks on German warships at sea or in naval dockyards. After the fall of Poland, all German interceptor fighters could be concentrated for the defence of such targets. The bombers were therefore liable to interception in overwhelming force.

Absence of self-sealing fuel tanks had made the Wellington very vulnerable to fire. One incendiary bullet striking a fuel tank could cause the destruction of a bomber. All

priorities for these tanks had rightly been given to fighters and few, if any, were as yet available for the bombers. Lastly, the fields of fire of the power-operated turrets of the Wellington proved to be deficient. It had been believed that the bombers were so fast that any effective fighter attack must come from astern. But the Germans, after studying the fields of fire of a shot-down Wellington, had realized that it was relatively defenceless against a beam attack. They therefore gave up attacks from astern and concentrated on beam attacks, accepting the large deflections of aim which such attacks involved.

The defects in the Wellington could be, and were, speedily put right. The fields of fire were modified to deal with beam attacks, and all our bombers were equipped with self-sealing tanks by the spring of 1940.

These early operations were, in themselves, of little importance, but they had a profound effect on our bombing policy for the next four years. The Air Staff concluded that Wellingtons and Hampdens would not be able to operate over Germany in daylight. The Whitley, slower and less well-armed, had always been regarded as a night-bomber; it

Wellingtons being refuelled

SHAPED FOR S T O L   Short take-off and landing (STOL) capability is one of the greatest assets an aeroplane can offer.   The French Dassault Mirage G prototype fighter (*above*) achieves it by use of a swing-wing.   DH Canada's Twin Otter, shown below in passenger and air survey form, has high-lift wings with full-span double-slotted flaps

1921: Fokker F.III of KLM. 240hp Armstrong Siddeley Puma engine. 5 passengers. Cruising passengers. Cruising speed 84mph

1929: Consolidated Commodore of Pan American. Two 525hp Pratt & Whitney Hornet engines. 32 passengers. Cruising speed 110mph

1930: Lockheed Vega of Braniff. 425hp Pratt & Whitney Wasp engine. 6 passengers. Cruising speed 150mph

1931: Curtiss Condor of Eastern Air Transport. Two 600hp Curtiss Conqueror engines. 18 passengers. Cruising speed 116mph

1931: Junkers G 24 of SHCA (Greece). Three 310hp Junkers L 5 engines. 9 passengers. Cruising speed 113mph

1932: Savoia-Marchetti S-66 of Ala Littoria. Three 750hp Fiat A24R engines. 14-18 passengers. Cruising speed 138mph

AIRLINE ANNIVERSARY 1919–1959

Just fifty years ago the world's first international
scheduled airline services were opened between
Britain and the Continent. No aircraft symbolises
better than the 360/490-seat Boeing 747 "jumbo jet"
(*above*) the immense industry that has grown from
those cross-Channel hops of 1919. With the supersonic Concorde, the 747 will be the flagship of the
'seventies, when more than 300 million people will
travel annually on scheduled services, some at more
than twice the speed of sound

1945: Lockheed 10A Electra of New Zealand National
Airways Corporation. Two 450hp Pratt & Whitney
Wasp Junior engines. 10 passengers. Cruising
speed 150mph

SEA KING
IN ACTION

One of the most vital tasks confronting the world's navies and maritime air forces is the detection, location and destruction of enemy submarines. This picture shows a Westland Sea King lowering its sonar into the sea to "listen" for submerged submarines. The hump on its back contains search radar. The Sea King's armament can include four homing torpedoes

was now decreed that the whole of our long-range bomber force should henceforth attack targets involving penetration of the German defences only by night. The light bombers, Blenheims and Battles, would continue to operate in daylight in support of the Field Force, protected by a general fighter cover.

Though this decision was taken somewhat hastily, before the defects in the Wellington had been remedied, subsequent experience showed that it was undoubtedly correct. The change of policy was possible because the Air Staff, while hoping that daylight operations would be feasible, had ensured that the aircraft were designed, and the crews trained, for night as well as day bombing— though it is true that the difficulties of night navigation, target-finding, and bomb-aiming had not been fully appreciated.

It is not generally realized how much the development of Bomber Command had suffered from the policies of successive Governments in the inter-war years. After the First World War there was a great surge of anti-war feeling, and our defence forces were cut to the bone. British Governments deluded themselves by thinking that if they disarmed others would follow suit. They were reluctant to build bombers, or any weapons that could be classed as "offensive", for fear of starting an arms race. They hoped that the Disarmament Conference, convened in 1932 at Geneva, would succeed in outlawing air bombardment. And in 1924 they had instituted the infamous "ten-year rule", which assumed that no major war could be expected for ten years. Unfortunately, each successive year was re-affirmed as the starting point of this tranquil period, so that it always remained at ten years.

In 1933 Hitler came to power in Germany, and forthwith embarked on a massive programme of re-armament. Then in 1934 the Disarmament Conference ended without

achieving any result whatever. The policy which had misguided successive British Governments since 1924 was now seen to be hopelessly discredited, and the years of illusion were over.

Our defence forces, which had been reduced to near impotence, had scarcely begun to recover by the time of the Munich crisis in September 1938, when our military weakness compelled the hapless Prime Minister, Neville Chamberlain, to make the best bargain he could with Hitler. The mild sense of alarm that had been felt in 1935, when two British Ministers, after visiting Germany, reported on the extent of Nazi re-armament in the air, was now powerfully re-inforced and made it inevitable that all priorities should be devoted to improvements in our air defence.

It was now realized that our re-armament had been too little and too late, and that we would have to endure an opening defensive phase in a war which seemed inevitable. Though this concentration on air defence was correct in the circumstances, it meant that the production and equipment of bombers was slowed down and even, in some instances, temporarily stopped. Thus it was that, when war came in 1939, Fighter Command was almost ready for its task, while Bomber Command was not.

At the outbreak of war we had 55 bomber squadrons, but this strength was illusory. The Government had vainly hoped that by putting everything "in the shop window" the Germans would be deterred. New squadrons, by no means fully manned and equipped were added to the battle order of Bomber Command, but we had no reserves of aircraft and no operational training organization.

When war came there was no longer any point in this piece of humbug, and urgent provision had to be made for war wastage in aircraft and crews. Ten squadrons of light bombers went to France with the Field Force, and ten more squadrons were withdrawn to act as operational training units. This left 35 squadrons in the front line, with a daily average of some 300 aircraft available for operations.

The main employment of Bomber Command was the dropping of propaganda leaflets, of almost incredible ineptitude and futility, on German cities and towns. Opposition, not surprisingly, was light, but the aircrews learned much about the difficulties of night navigation in wartime. Air Staff requirements for radar aids for night bombers had been issued in 1938, but the over-riding demands of air defence had prevented any substantial progress.

Hitler's assault in the west, which began on May 10th, 1940, involved the light bombers in extremely intensive daylight operations. They suffered very severe casualties in a hopeless effort to avert the fall of France. These losses were serious, as they consisted mainly of highly-trained experienced aircrews; the seed-corn of future expansion. The night bombers attacked German communications and the factories of the Ruhr. These attacks were not very effective, and could not halt the German advance against the rapidly disintegrating French armies and air forces.

The strategic concept with which the allies had begun the war lay in ruins all around them. France, Holland, Belgium, Denmark, Norway, and the western half of Poland had fallen to Hitler's assault, and eastern Poland had been absorbed into the maw of Russian Communism. Even the cause for which we had entered the war, the integrity of Poland, had vanished. We now stood alone, and all that we could do was to hold on as best we could.

The important part played by Bomber Command in the Battle of Britain is not always fully realized. Our object in the battle was not only to parry the German attacks and maintain air superiority in our own skies, but to defeat the German invasion plan—Operation "Sealion". Night after night the bombers pounded the Channel ports, in which large numbers of troops and a great quantity of stores, with a vast array of huge barges, had been assembled. These attacks were remarkably successful and immense damage was done. For example, on the night of September 13th, a very effective attack sank no fewer than eighty great barges in the port of Ostend alone.

These attacks and, of course, the very heavy casualties inflicted on the German air force by Fighter Command, caused Hitler on September 17th to postpone Operation "Sealion" indefinitely. Orders were given to move the troops and the remaining barges and equipment away from the dangerous vicinity of the ports.

The immediate threat of invasion had passed, and it was now our task to do what we could to reduce the military and industrial strength of Germany. Early in September when the Battle of Britain was approaching its climax the Prime Minister, Winston Churchill, had written a masterly minute for his colleagues in the War Cabinet. In this he said "The Navy can lose us the war, but only the Air Force can win it. Therefore our supreme effort must be to gain overwhelming mastery in the air. The fighters are our salvation, but the bombers alone provide the means of victory. We must, therefore, develop the power to carry an ever-increasing volume of explosives to Germany, so as to pulverize the entire industry and scientific structure on which the war effort and economic life of the enemy depend, while holding him at arm's length from our Island. In no other way at present visible can we hope to overcome the immense military power of Germany . . . . "

The prospects, at this time, of any worthwhile offensive air action against Germany were rather bleak, and several serious difficulties would have to be overcome before such a policy could be made effective. Two

Hampden in service with No 185 Squadron during the early months of World War II

Below: Whitley V, code-lettered EY-W, of No 78 Sqn. Each bomb painted on the nose signifies an operational sortie

Blenheim IVs of No 40 Squadron, employed on hazardous daylight raids in the Summer of 1940

of these difficulties were soluble, given time and a correct re-allocation of priorities, but the third was inherent in the strategic situation.

First of all, Bomber Command was much too small for its gigantic task. Secondly, it had not yet been possible to provide the radar aids to night-bombing that would be essential to success. Top priorities in research, development, production, and training would have to be switched to the requirements of Bomber Command. Lastly, the bombers were faced by an entirely new strategic situation. Successive Governments had based all pre-war planning on the assumption that our bombers would be able to use advanced bases on the Continent, and the penetration needed to reach the Ruhr and other targets in western Germany would therefore be relatively small. Almost the whole of Germany would be within reach of the night bombers, even during the short nights of summer.

Before the war it was no doubt politically impossible to plan on any other assumption, but as a consequence the Air Staff had never been allowed to make preparations for the

situation in which we now found ourselves. It was, therefore, broadly speaking true that our bombers were not specifically designed and equipped to perform the tasks now expected of them.

It was in these circumstances that in October 1940 the War Cabinet ordered the heaviest possible air offensive against Germany. It was soon apparant that, without radar aids, our aircrews, no matter how hard they tried, were unable to identify the small military targets that they were expected to attack, especially in densely built-up areas such as the Ruhr. The German defences were as yet comparatively ineffective, and it was common for bombers to spend half-an-hour or more cruising around trying to find their targets. Even so, it was becoming increasingly obvious that only a small proportion were able to find their objectives. Targets with some unmistakable feature, such as a large estuary or lake, proved to be the easiest to find.

Throughout 1941 the Command struggled to carry out its task, with insufficient numbers and inadequate equipment. It was a year of frustrated effort and hope deferred. There had been no effective shift of priorities to the requirements of the air offensive. Indeed our mounting losses at sea, and the painful inability of the Royal Navy to cope with the submarine menace, caused the Prime Minister, on March 6th, to give absolute priority to the Battle of the Atlantic. Additional squadrons of long-range aircraft for anti-submarine warfare could be found from only one source, and during 1941 no fewer than seventeen squadrons (204 aircraft with crews) were transferred to coastal Command. They were supposed to be on loan, but none of them ever returned.

In addition, a considerable number of aircraft with crews were transferred to the North African theatre. As a result Bomber Command, which had succeeded in forming some nineteen new squadrons during the year, lost to Coastal Command and the Middle East a force of equivalent size, ending a year of great effort with no increase in strength, indeed temporarily weakened by casualties and its successful efforts to expand.

The War Cabinet had failed to make the redeployment of the national effort on which the success of the air offensive depended. There had been but little increase in the strength of Bomber Command, and it had become quite clear that the selection and destruction of the precise military targets, on which the Government continued to insist, was impossible.

Towards the end of 1941 HQ Bomber Command prepared an appreciation of the situation. A study of raid reports and of photographs of aiming points, taken by the light of flash bombs at the moment of bomb release, showed that only a small percentage of our attacks reached their prescribed targets. Of course, in an area such as the Ruhr or a large industrial town, most of the bombs would cause some sort of damage to houses, roads, gas, electricity and water mains, or indeed to anything that they chanced to hit. But it was dishonest to go on claiming that the bombers attacked only specified military targets.

An analysis of the German night bombing of London showed that the density of bombing, over an area of 225 square miles centred on Charing Cross Station, was remarkably constant. Approximately the same number of bombs per square mile fell in the outer suburbs as in the central parts of the city or the dock areas. This even distribution was characteristic of night bombing by individual aircraft using normal methods of navigation and visual identification of targets. Bomber Command had little doubt that an analysis of its own attacks on the Ruhr, supposed to be aimed at such things as key factories and synthetic oil plants, would show a very similar pattern of distribution.

The solution of the problems of night-bombing was the provision of effective radar aids to navigation, target identification and bomb-aiming. But these aids were not immediately available, though enough work had been done to show that in time they could be provided.

In the meantime, the Command proposed that a more realistic policy should be adopted. Our experience of the German night-bombing attacks had shown that our heavy losses of industrial output had been due, not so much to the destruction of individual factories, as to the dislocation caused by the cutting off of supplies of electricity, gas and water, damage to communications, and above all to absenteeism caused by the destruction of workers' dwellings. They also showed that, weight for weight, incendiary bombs were far more effective in industrial areas than high explosives.

The effect of an HE bomb is limited to the direct effect of its blast and fragmentation, whereas the incendiary bomb can take advantage of the combustibility of the target. Thus one incendiary bomb might, for instance, set fire to a large building and bring about its total destruction. Very few objects are resistant to incendiary attack, and the most unpromising things will burn.

The policy advocated, therefore, was to attack large industrial cities and areas, using a high proportion of incendiary bombs. At some time in the future, the new techniques made available by radar devices would make it possible to select and destroy specific targets of especial importance.

This new policy, the so-called and much abused "area bombing", was the only one practicable in the circumstances. It did not, as is so often asserted, involve an indiscriminate scattering of bombs in residential areas. On the contrary, the most concentrated pattern of bombs of which the Command was capable, using a precise aiming point such as a railway station or a marshalling yard, was too big to be contained within any but the largest industrial town.

The increasing effectiveness of the German air defences brought about a revision of Bomber Command's tactics. The old "go as you please" methods were no longer practicable. In order to saturate the defences and overwhelm the fire-fighting organization, attacks were increasingly concentrated in time.

The arrival in substantial numbers of the four-engined heavy bombers, of which production had been shelved during our pre-occupation with air defence, was now greatly increasing the hitting power of the Command. By the time Air Marshal Sir Arthur Harris arrived, in February 1942, to take command, many of the worst of the bombers' teething troubles were over. It is time that no radar aids except "Gee"—an aid to navigation—were as yet available, but new and better devices were just round the corner.

In sixteen months of unremitting effort, the Command had inflicted substantial damage on German industry and naval installations, and had achieved a considerable measure of expansion, though this had been largely offset by transfers to Coastal Command and the Mediterranean campaign. Its rate of loss, though grievous, had been kept to a tolerable level. But those who disbelieved in, or disapproved of, the bombing campaign, lost no opportunity of pointing to the admitted lack of accuracy of night-bombing, and urging that the Command should be diverted to other tasks, such as the purely defensive war at sea, or the support of our land forces in North Africa, in a subsidiary theatre where the war could neither be lost nor won.

Harris was convinced that if he could attack an important industrial target with a great force of bombers, even if once only, it would show the Government—and the world—what a properly equipped and expanded bomber force could do. This idea culminated in the first 1,000-bomber attack on Cologne in May 1942, a spectacular project which aroused tremendous enthusiasm and raised the morale of the Command to new heights.

The results of this attack were dramatic. Losses were only 44 out of 1,046 aircraft despatched, the damage was immeasurable by any previous standards, and the Prime Minister was greatly impressed. He gave orders which, at last, ensured that Bomber Command would get the priorities it so desperately needed.

In due course there followed complete re-equipment with heavy bombers, effective radar aids, and the formation of the Pathfinder Force. The courage and determination of the bomber crews had brought a long period of difficulty and frustration to a triumphant conclusion.

The foundations were now well and truly laid for the great bomber offensive which, in conjunction with the American heavy bombers operating in daylight, created the conditions for the successful invasion of the Continent in June 1944.

The Luftwaffe's "Tradition Squadrons"

by André Van Haute

WHEN, on April 21st, 1961, the then C. in C. of the *Bundes-Luftwaffe*, Lt Gen Jozef Kammhuber, presented armsleeves to the *Geschwadernkommodoren* of JG.71, Jabo.G.31 and AG.51, he entrusted them and all the members of these units to continue the traditions established by World War I aces Manfred von Richthofen, Oswald Boelcke and Max Immelmann. There had been an intermediate link, as the former *Luftwaffe* of World War II had continued the traditions set up in 1914-18.

Immelmann:

Born in Dresden in 1890, Max Immelmann joined the Dresdener Kadetcorps at the age of 15, and left the army after obtaining his commission in 1912. He wanted to study engineering at the university, but the outbreak of World War I put an end to this plan and he was recalled to active service, joining the *Fliegerersatzabteilung*.

As is often the case with future aces, he was far from a good pilot at the start and suffered many forced landings. However, with courage and perseverance he managed to pass his *Feldpilotenprufung* in March 1915 and joined *Feldfliegerabteilung* No 62 in the Champagne region of France in April.

In May this unit moved to northern France on reconnaissance duties. Among its pilots

was a certain Oswald Boelcke, who had been allocated the first "C" type aircraft, made especially to protect unarmed reconnaissance machines. Immelmann became one of Boelcke's friends and it was only natural that after the latter got a new Fokker fighter, the "C" type machine went to Max Immelmann.

From then onwards there was a keen rivalry between the two friends which lasted until Immelmann's death a year later. For his eight victories, achieved mainly on Fokker monoplanes, he received Germany's highest award, the "Pour le Mérite", in January 1916. A handwritten letter from Emperor Wilhelm II congratulated him on his 13th victory in March, but his luck ended on June 18th. Before he even joined in the fight, his aircraft was seen disintegrating, probably due to structural failure, and he crashed not far from Lille.

His death caused great dismay in Germany, and even his enemies recognised his chivalry in combat. Some days after his death, a British aircraft dropped a wreath and a letter over the lines, stating that Lt Immelmann had been considered a gentlemen by all British fliers—certainly no mean compliment in the middle of so dreadful a war.

On the creation of the new *Luftwaffe* in the early thirties, it was decided to grant the

The Fokker monoplane was the first real fighter of the 1914-18 War, with a machine-gun synchronised to fire between the propeller blades. Max Immelmann was the greatest of its pilots

name "Immelmann" to a crack unit; so *Stukageschwader* No 163, with HQ at Insterburg in East Prussia, was honoured with this title. The *Gruppen* of this unit were dispersed, in Wertheim/Main and Altenburg/Thüringen. During the Spanish Civil War, a flight of its Ju 87A dive-bombers was detached to assist the Spanish Nationalists in December 1937. The operational capabilities of the "Stuka", as the Ju 87 was known, were thus tested in combat and some wrong conclusions were drawn, with disastrous results in another battle to come . . . .

St.G.163 was renumbered St.G.2 at the start of the war with Poland, where, in company with other Stuka units, it bore the brunt of the first onslaught. On May 10th, 1940, as part of *Fliegerkorps* VIII, it took part in the battle of the Low Countries, bringing destruction and terror to soldiers unaccustomed to this kind of warfare. However, far from being a wonder weapon, the Ju 87 was easy meat for modern fighters flown by determined pilots, and during the opening stages of the Battle of Britain, the Immelmann units encountered such stiff opposition that the *Luftwaffe* high command had to withdraw them temporarily. Flying Ju 87B's, No II *Gruppe* moved to Sicily in January 1941, whilst Nos I and III *Gruppen*, which had stayed behind in France, took part in the Balkan campaign in April of the same year. Later they saw action against Crete, as part of *Luftflotte* 4.

On June 22nd, 1941, Operation "Barbarossa", the war against the Soviet Union, started and *Fliegerkorps* VIII included two *Gruppen* of St.G.2. One of their young pilots, named Hans Ulrich Rudel, made a name for himself by sinking the Russian battleship *Marat* in Kronstadt harbour. He went on to become the best-known *Stuka* exponent on the Russian front, and it was only at the very end of the war, when *Schlachtgeschwader* 2 (as St.G.2 had been known since 1942) and other *Stuka* units re-equipped with new

aircraft, that he gave up his Ju 87G, although S.G.2 pilots normally flew the Ju 87D variant. The re-equipment programme for the *Schlachtgeschwadern* had been initiated at the end of 1943 and S.G.2 ended the war with an attack version of the Focke Wulf Fw 190A and the remaining Ju 87G tankbusters.

Almost 14 years after the collapse of Hitler's Third Reich, a new "Immelmann" unit was born, at Erding in western Germany.

It was, in fact, on April 1st, 1959 that *Waffenschule* 50 (equivalent to an RAF OCU) was given the task of forming Tactical Reconnaissance Wing No 51 of the *Bundesluftwaffe*. Its aircraft were sleek-looking Republic RF-84F Thunderflashes. On July 7th, the 36 aircraft of *Aufklarungsgeschwader* AG.51 and 1,200 men of this unit were paraded for the official handing-over ceremony to their first Kommodore, Oberstleutnant Grasemann.

On May 5th, 1960, AG.51 moved to the newly-opened base at Manching. An aerodrome had existed there since 1936 and during World War II had housed Me 110 and Ju 88 night fighters. It is situated 5 miles south of the town of Ingolstadt/Donau, at a place which already existed in 100 BC and was then known by the Celtic name of Oppidum.

On August 1st, 1962, AG.51 was officially assigned to NATO, under the 4th ATAF, as part of the 5th *Luftwaffen* Division. Oberstleutnant Loosen took over on December 5th of the same year. 1963 was probably "the year" for the Immelmänner, as they started to re-equip with the RF-104G Starfighter. Pilots and ground crew had to work hard to cope with the new highly supersonic aircraft and it was not until June 1965 that the last RF-84F flight took place, AG.51 having by then got its full complement of Starfighters.

Boelcke :

Oswald Boelcke, another of the German aces of World War I, was a different type of man to his friend Max Immelmann. Fair but strict, he was not the sort of "young fighter pilot" future generations learned to expect. Another unusual fact was that he flew his early reconnaissance missions with his older brother Wilhelm as his observer. At the end of 1914 they were concerned mainly with artillery spotting and Wilhelm, who had received his training at the War Academy, made a good team with Oswald, who was a keen pilot. However, Oswald soon became impatient at having to make unarmed reconnaissance missions.

Drastic changes in the air arm's policy took place when a Major Thomsen reorganised the whole *Fliegertruppen*. *Fliegerabteilung* 62, to which Boelcke belonged, was the first to receive one of the new "C" type aircraft. which had a machine-gun in the rear cockpit; in company with his gunner, Husarenleutnant von Wühlisch, he achieved his first victory in air combat on July 4th, 1915.

The "C" type aircraft was heavy and sluggish on the controls, and only above average pilots had any success with it. However, Anthony Fokker's design office had developed the "E" type single-seat monoplane fighter, armed with a synchronized machine-gun firing between the propeller blades. This, the first real fighter aircraft of the *Fliegertruppen* was superior to anything the Allies had at that moment. When Fokker demonstrated it at Douai in June 1915, young Boelcke showed at once a great urge to fly it in combat, so *Fliegerabteilung* 62's CO, Hauptmann Kastner, gave him permission to do so.

The High Command, realising the superiority of the "E" type, decreed that only defensive patrols over the German lines were to be flown, to prevent an example of the aircraft and its synchronising gear from falling into Allied hands. This was contrary to Boelcke's belief that one had to go after the enemy and fight him—a tactic that led

to use of the term "*Jagen*" (to hunt) and hence the words *Jagdflieger* (fighter pilot) and *Jagdstaffeln* (fighter squadrons).

After his eighth victory he received the coveted "Pour le Mérite" award, at the same time as his friend Max Immelmann. After his 19th victory he was retired from combat duties on the orders of the Kaiser himself. This was a direct result of Immelmann's death and the wish to take advantage of the experience of one of Germany's best pilots in teaching tactics to the younger generation.

Against his own wish, Boelcke was sent on an instructional tour which took him to the Balkans, Turkey and the eastern front. But when the Allies launched their Somme offensive in July and August 1916, their overwhelming air superiority took the Germans completely unawares and they had no idea what tactics to use against the large enemy formations. On August 12th, Boelcke was still in the east when an urgent order from the *Feldflugchef* requested his immediate return to the western front, to form *Jagdstaffel* No 2.

Before leaving, and on the recommendation of his older brother, he recruited Ulanenleutnant Manfred Freiherr von Richthofen and the much older Leutnant Erwin Böhme. The former became his most illustrious pupil, whilst Böhme became one of his closest comrades and later commanded the *Jagdstaffel* "Boelcke".

On the Somme front, Boelcke and his whole *Staffel* quickly built up their totals of combat successes, due in part to availability of the new Fokker "D" type biplane and also to flying as a complete *Staffel* instead of operating single sorties. On October 26th Boelcke gained his 40th victory; the air arm's chief, Gen von Hoeppner, called his squadron "the leading one" and urged other units to follow its example. Two days later, Captain Boelcke was killed, at the age of 25. By order of the Emperor *Jasta* 2 became known as "*Jagdstaffel* Boelcke".

Unarmed Arado Ar 65s were used for operational training by the reformed Richthofen *Geschwader* in 1935

With a maximum speed of 205mph and armament of two machine-guns, the He 51 equipped the first fighter squadrons of Hitler's *Luftwaffe*

The next "Boelcke" unit was formed in 1935, at Fassberg, under the later General der Flieger Richard Putzier. Designated *Kampfgeschwader* KG.27, it was equipped from the start with the new Heinkel He 111 bomber, and flew successive versions of this aircraft right up to the end of the war.

In 1936, the *Geschwaderstab*, and *Stabsstaffel* (headquarters unit) were stationed at Hannover Langenhagen, with No 1 *Gruppe*, whilst No II *Gruppe* went to Wunstorf/Hannover and No III *Gruppe* to Delmenhorst/Bremen. KG.27 took part in the campaigns in Poland, France, and in the Battle of Britain, in which it suffered severe losses, probably due to vulnerability of the He 111 crew compartment and lack of fighter escort. After the

Battle it was moved to the eastern front, where it was used mainly in direct support of the infantry instead of in the strategic role for which it was intended.

There was, however, at least one operation in which it was used in the strategic role, in company with other He 111 units. This was the famous raid on Poltava, Pirjatin and Migorod in June 1944. Taking off from Poland, the aircraft set out to destroy in particular American 15th AAF B-17 Fortresses and P-51 Mustangs. The raid caught the Russians and Americans completely off guard, over 50 US aircraft, some Soviet

Members of *Jasta* 11 wait to be presented to General Ludendorff by von Richthofen, in Belgium, August 19th, 1917

types and the units' fuel dumps being completely destroyed. This *coup d'eclat* represented virtually the end of KG.27 as a strategic bomber unit; its last Kommodore was Oberst Freiherr von Beust.

The third of the "Boelcke" units was formed from the nucleus of *Waffenschule* 50 at Erding, which, incidentally, was the very first unit to equip with the Republic F-84F Thunderstreak fighter-bomber. *Jagdbombergeschwader* 31 "Boelcke" was formed at Büchel on September 1st, 1957, but moved later to Nörvenich. Its first CO was the then Major Gerd Barkhorn, a former Kommodore of JG.6 during World War II. By June 20th, 1958 the unit was fully operational and its No I *Gruppe* went to Bandirma in Turkey for two months' intensive gunnery and bombing training. On July 1st, 1961, Lt Gen Harlinghausen, of *Luftwaffengruppe Nord*, handed badges bearing the name "Boelcke" to members of *Jabo*G.31, in the presence of relatives of the late Oswald Boelcke.

Re-equipment with F-104G Starfighters started that same year, some of the pilots making their conversion to the new type whilst others were transferred with the remaining F-84F's to Decimomanny in Sardinia. Oberst Barkhorn left in 1963 and was succeeded by Obstlt. Meyn, who in turn handed over command to Oberst Obleser on January 2nd 1964. The *Geschwader* had the honour to form the German contingent in the fly-past to celebrate the 41st anniversary of the Turkish Republic, over Ankara on October 29th, 1963. It now forms part of the 2nd ATAF and is an important component of the NATO strike force.

Von Richthofen:

As already noted, it was Oswald Boelcke who brought Manfred von Richthofen to the western front as a member of *Jasta*.2. He soon became Boelcke's best pupil and on September 17th 1916 took part in his first operation with the complete *Staffel*, headed by Boelcke. There and then he gained his first victory, followed by many others in a very short period. With 16 successes to his credit, he was promoted CO of *Jagdstaffel* 11 and was awarded the "Pour le Mérite".

Boelcke's example and teaching had instilled in him a clear and calculated combat technique and he was not the reckless type of fighter pilot. His planning and foresight enabled him to surpass quickly his former master's tally of combat victories and eventually to double it.

In the spring and summer of 1917, the Allies made a series of concentrated attacks and there was an acute shortage of German fighters to counteract these. It was decided, therefore, to form large formations of fighters to throw into the battle wherever and whenever it was necessary to gain local air superiority. It was on 25-year-old Rittmeister Manfred von Richthofen that the commander of the *Luftstreitkrafte* bestowed command of the first *Geschwader*, with about 30 aircraft on its strength.

J.G.1 was formed officially on June 23rd, 1917 from *Jagdstaffeln* 4, 6, 10 and 11. Richthofen recruited personally most of its pilots, from squadrons and fighter training schools, and his choice must have been sound as J.G.I scored its 100th victory after a bare two months' existence.

Richthofen, or the Red Knight as he was to be known, always led his *Geschwader* and acheived his own 80th victory in April 1918. His end came on the 21st of that month, when, according to witnesses, he was shot down by Capt Roy Brown, flying a Sopwith Camel. He fell in the Allied lines and the British gave him a burial with full military honours. Three weeks after his death, his *Geschwader* received the name "*Jagdgeschwader* Freiherr von Richthofen" No 1.

His body was brought back to Germany after the war and he was buried at the Invalidenfriedhof in Berlin on Nov 21st, 1925. Nine years later the fledgling *Luftwaffe* was

Rittmeister Manfred von Richthofen, officially the most successful fighter pilot of the 1914-18 War, with 80 confirmed victories

forming a new fighter arm and April 1934 saw the birth of the *Reklamestaffel Mitteldeutschland e.V* at Döberitz aerodrome. The *Staffel's* leader was Oberstleutnant Ritter von Greim, who achieved 25 confirmed victories in World War I and wore the "Pour le Mérite". By a coincidence he was also to be the last Chief of Staff of the *Luftwaffe* in 1945.

After Hitler's public revelation of the defence act of March 13th, 1935, the secret *Luftwaffe* could appear at last in its official guise. The Air Minister, General der Flieger Hermann Göring, himself the last Kommodore of J.G.I Richthofen, called the great ace's deeds an example for the *Reichsluftwaffe* to follow, and the first *Geschwader* was named J.G.No. 132 Richthofen.

The first *Gruppe* of J.G.132 was set up at Döberitz, and II *Gruppe* was added, at Jüterborg-Damm, on April 1st, 1935, under Major Raithel. Major von Doering, a former leader of Jasta.4, was the Kommandeur of No 1 *Gruppe*. A keen spirit of competition soon developed in the two *Gruppen* and the Arado 65, which had been used for initial operational training, was soon replaced by the Heinkel He 51, the first real fighter aircraft to serve the new *Luftwaffe*. J.G.132 was chosen then to "show the flag" in various parades and to perform before the many foreign visitors who came to Döberitz.

With the rapid development of the *Luftwaffe* in 1936/37, J.G.132 Richthofen became the parent unit of other newly-formed fighter *Geschwadern*. For example J.G.133, with HQ at Wiesbaden, received its ground crews from J.G.132.

In 1937, J.G.132 provided the air defence force for Berlin during the large-scale *Wehrmacht* manoeuvres, equipped with the new Messerschmitt Bf 109B. On March 12th, Hitler started the annexation of Austria and its No I *Gruppe* transferred to Wien-Aspern; it was, however, back in Germany after a week. April 21st, 1938, was a big day in the *Geschwadern's* history, for the twentieth anniversary of the Red Knight's death saw the inauguration of a Richthofen monument at the entrance to the Döberitz fighter station.

A change in policy concerning *Geschwadern* numerals resulted in its number being changed from 132 to J.G.2 "Richthofen" in 1939.

During the Sudeten crisis J.G.2 stayed in the Reich, but a 10th *Staffel* had been formed in the meantime, as the II *Gruppen* had departed to form the nucleus of *Zerstorergeschwader* I, with Bf 110's, in 1938. The duty of J.G.2 was still to protect Berlin, so it did not take part in the Polish campaign, although the 1st *Staffel* did go to Lyck, in East Prussia, were it flew security patrols.

On November 2nd, 1939, the 40 aircraft of the *Geschwader* transferred to Frankfurt-Main, to provide fighter escort for Dornier

Crews of KG.27 "Boelcke" lined up in front of their
He III twin-engined bombers

End of the road for one of KG.27's He IIIs, which
crashed on March 29th, 1940

Do 17 reconnaissance aircraft, and on
November 22nd contact was made with the
enemy for the first time. Lt Wick and Obf.
Kley each shot down a Curtiss Hawk of the
Armée de l' Air.

At the start of the attacks on Holland and
Belgium, the unit had its II Gruppe stationed
in central Germany. No III Gruppe, under
Major Dr. Mix, together with I Gruppe,
operated in the Metz-Maubeuge area. No
II Gruppe saw operations in Holland and
Belgium under Fliegerkorps II.

Major Mix was shot down on May 21st
after having destroyed a Morane MS.406,
but managed to regain the German lines.

Before the end of the Battle of France the
Geschwader moved to Evreux, but I and II
Gruppen of J.G.2 moved later (June 23-27th,
1940) to Beaumont le Roger, which had been
an RAF base during the earlier months of
the war.

No III Gruppe of J.G.2 had a short spell of
rest in Frankfurt before being posted to Le
Havre. During the Battle of Britain there
had been fierce competition between the
various JG's and the "Richthofen boys"
played their part in this, Oesau, Balthazar
and Wick being among the more successful
pilots. Major Wick became Kommodore
after the Battle, but failed to return from a

Lt Max Immelmann, known as the "Eagle of Lille" and highly respected by friends and enemies alike

First aircraft flown by the third-generation "Richthofen" pilots were these Canadair Sabre VIs

sortie on November 28th, 1940, after his 56th victory.

After the Battle of Britain, and in the face of increasingly aggressive RAF fighter sweeps, the *Gruppen* dispersed to Normandy Bretagne (Morlaix) and a 10th *Staffel* (Jabo) was added for nuisance raids on British coastal towns.

During the Channel dash of the German battleships *Scharnhorst* and *Gneisenau*, the *Geschwader* moved to Coxyde in Belgium; but after RAF raids on St Nazaire and elsewhere, Jafu.West ordered further dispersion inland. During May 1941, the I *Staffel* had received Bf 109G's and became No 11 *Staffel*.

By May 1942 the whole *Geschwader* was re-equipped with the Focke Wulf Fw 190A, except for No 11 *Staffel* which kept its Bf 109G's.

After the Allied landings in North Africa, No I *Gruppe* of J.G.2 moved to Marseille Marignan and Bourges; No III *Gruppe* went to Istres, whilst II *Gruppe* and the

eleventh *Staffel* went to St Pietro in Sicily. By the start of 1943, all units were back in northern France, were they fought heavy engagements with the RAF and USAAF. Major Oesau, Kommodore during this period, left on June 31st, having been with the unit since 1936.

After the Allied invasion, the story of J.G.2 followed closely that of the retreating German forces. Its last Kommodore was Oberst Bühlingen, who led it from May 1st, 1944 until the final day of the war.

It was only befitting that on the creation of the new *Luftwaffe* in the mid-fifties one of its units, J.G.71, was entrusted to carry on the traditions of Manfred van Richthofen. It first flew Canadair Sabre VI's, under Oberstleutnant Erich Hartmann, the top scoring *Luftwaffe* pilot of World War II, its base being at Ahlhorn. In 1962, the Sabres were gradually replaced by F-104G Starfighters. The unit had moved to Wittmundhafen in the meantime and was taken over by Oberst Josten on May 29th, 1962. He had the task of getting J.G.71 operational on the F-104G, and finally left on March 31st, 1967, handing over to Oberstleutnant Horst Dieter Kallerhof. As J.G.71 and J.G.74 are the only *Luftwaffe* interceptor units, the task of their present Kommodore is certainly as vital as that of any of his illustrious predecessors.

Only interceptors in the present *Luftwaffe* are the F-104G Starfighters of the "Richthofen" squadrons

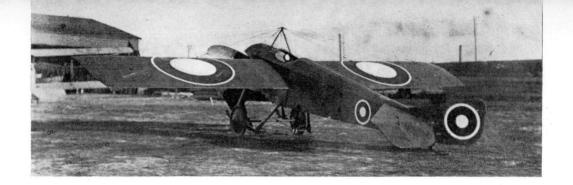

Coastal Colours

by Bruce Robertson

MOST military aircraft are camouflaged and their colours relate to the environment in which they operate, land or sea. In the case of coastal aircraft a compromise is necessary, with operational considerations tipping the scales either for concealment at their land bases, or while operating over the sea.

In the First World War, up to April 1st, 1918, British coastal aircraft were under the control of the RNAS and no distinction was made between aircraft of the Fleet and those on coastal duties. In general, they were not camouflaged up to 1916; subsequently, upper and side surfaces were covered with pigmented dope to specification PC10 or PC12 which rendered them either dark green or chocolate brown. This was largely for protective purposes, to prevent the fabric deteriorating through harmful rays from the sun. It was the exceptions that provided the more colourful schemes.

The Admiralty made plans in mid-1917 for photographing the Kiel Canal, for naval intelligence purposes. In July two D.H.4's, A7457 and A7459 with 200hp RAF3A engines, were allotted to the task and fitted with extra tankage to give 14 hours endurance. At Hendon they were given a special camouflage of sky blue and biscuit, which was then thought to ensure minimum visibility against a clear sky—conditions essential to success. However the operation was called off and the two aircraft, retaining their camouflage, operated from Bacton on anti-submarine work. One was lost at sea on September 5th, 1917 and its crew were rescued by a Curtiss H.12 flying-boat.

These flying-boats, together with the later F.2A and F.3 types, were used largely for anti-submarine patrol. In general their finish was in the dark green/brown standard pigmented cellulose, but some crews favoured the most gaudy of striped colour schemes. One aircraft, N4289, was described as "terrible in appearance, painted post-box red with yellow lightning marks running diagonally across". The pilot, Captain G. F. Hodson of Yarmouth, said he hoped it would put the wind up the enemy, but such markings were really permitted for a more practical reason, in that it helped the sighting of machines forced to come down at sea. At Yarmouth Air Station the crews vied with each other to produce the most startling schemes, in striping and dicing.

Different from the start—a coastal aircraft of 1915, a Dyott monoplane, at Dover with the RNAS red ring insignia. Not until November 1915 did the RNAS adopt red, white and blue roundels in common with the RFC

40

Fawn and blue—one of two specially camouflaged D.H.4s intended for bombing the Kiel Canal. Even the blue and white of the rudder stripes were reversed

Perhaps the most colourful of coastal aircraft in World War I were the flying-boats operating against each other across the Adriatic. The Austrians displayed their national colours of red, white and red stripes at the wingtips, and the Italians followed suit with their red, white and green. At a distance, when shades could not be discerned but only the striping, identification was difficult—doubly so, as both Italy and Austria used flying-boats of basic Lohner design. The Austrians therefore added the German *Cross Patée* (colloquially, the Iron Cross inboard of the stripes; for good measure they also painted the hull noses white and added a cross there. In similar positions, the Italians then added their colourful red, white and green roundels, but omitted the striping to differentiate further from the embellished Austrian craft.

The Germans, in general, used plain colours. A sea grey camouflage, with national and serial markings in black and white, was normal, but a few German seaplanes did have the fascinating camouflage of lozenges of different colours. This scheme was exclusive to Germany and its allies, simply achieved by the use of printed fabrics.

Between the wars, when fighter aircraft were embellished in gaudy squadron schemes, flying-boats in the RAF were standardised with seaplane grey enamelled hulls and aluminium superstructures. When camouflage was introduced in 1937, coastal aircraft were not immediately affected; and when war came, the landplane colours of green and brown were applied. Not until 1940 was the Temperate Sea Scheme devised, with a disruptive pattern of dark slate grey and extra dark sea grey for the upper and side surfaces of Coastal aircraft.

During the war, the familiar red, white and blue roundels were changed to just red and blue on the upper surfaces of RAF and Fleet Air Arm aircraft, so that the white did not stand out and compromise the camouflage—

Austrian colours and German crosses; an Austro-Hungarian Lohner Type L flying-boat in 1916

A Felixstowe flying-boat striped to assist location and identification at sea, in 1918

A curious Italian embellishment—a wound stripe for each bullet hole sustained by the aircraft, in this case Macchi L.3 (redesignated M.3) No 7347

except on Coastal aircraft. This exception resulted from several attacks by Spitfires and Hurricanes on Coastal aircraft—Blenheims being mistaken for Ju 88's and Hampdens and Hudsons for Do 17's—culminating in the tragic shooting down of a Coastal aircraft in October 1939, after which the order went out to the Command to revert to the normal roundel on all surfaces.

Early in 1943 came the most radical change of all in coastal aircraft colours—the new white finish. In early 1942, with the Battle of the Atlantic at its height, Coastal Command's main effort was devoted to anti-submarine operations. A scientist on operational research, when analysing crews' reports, became aware that few bombing or depth-charging attacks on U-boats were effective after they had submerged, even if they could be seen under the surface.

Before the introduction of the snorkel "breathing device" U-boats were forced to surface for a few hours every day. Once they sighted a patrolling aircraft, it took some 45 seconds to disappear beneath the surface, where they were normally invisible to aircraft not directly overhead. Scientists thus directed their efforts to making the aircraft less visible to U-boat crews, estimating that all-white aircraft would increase by 30 per cent the difficulties of detection and so allow an aircraft to approach the submarine without being spotted. In effect, the scientists calculated, an enemy submarine which submerged after sighting a dark aircraft would have been on the surface for twelve seconds longer if the aircraft had been white.

As a full-scale experiment, two Whitleys of No 502 Squadron were painted off-white late in 1941. Towards the end of the same year, observers at Limavady, equipped with stop-watches, did timed observed approaches of alternating white and black painted Wellingtons. To obtain more precise measurements under controllable conditions, the General Electric Company were called in.

They experimented with three model Wellingtons, one black, one white and one in the prevailing temperate sea scheme of sea shades on all except the undersurfaces which were duck-egg blue. Their conclusion was that, by day and night and in simulated moonlight, a white finish offered a 10 per cent advantage over all others, except at half-moon when the differences were negligible.

An all-white scheme was recommended and adopted for aircraft operating in areas where enemy aircraft were not likely to be met, such as in the Azores. The first operational aircraft to bear the new scheme were those of No 53 Squadron at St Eval, in June 1942, when they were preparing to move to Quonset Point in the United States and to operate from Rhode Island, in order to search waters at the very gateway to New York harbour then being threatened by German U-boats.

It was a different matter for coastal aircraft based in the United Kingdom, with its south-east corner a mere 21 miles from German-occupied territory. Conspicuous all-white aircraft dispersed on airfields might invite attack. However, such was the menace of the U-boat attacks that the marginal advantage of gaining 12 seconds when attacking enemy submarines on the surface took precedence over their concealment from bombing attack. One compromise was that the strict plan view would be extra dark sea grey for aircraft operating from the UK; but the complete sides, with the large fin and rudder areas, were, from February 1943, painted an unvarying white, setting a style for years to come.

In the same way that some Tiger Moths had been fitted with bomb-racks when Britain was threatened with invasion in 1940, so in 1943 it was rumoured that they were to be used for coastal anti-submarine spotting in the same way as D.H.6 trainers had been used in 1918. These rumours spread from the sighting of an all-white Tiger Moth in

December 1943. It was, however, the result of a prank, with the very best of intentions, by Canadian airmen. No 421 (RCAF) Squadron personnel had invited children living near Tangmere to a Christmas party, at which the climax was the arrival of Santa Claus, with a bag of toys, in an all-white Tiger Moth named *North Pole Express*.

When "Invasion Stripes" were used to identify aircraft during the Normandy invasion on June 6th, 1944, Coastal Command aircraft in general were not affected; but for a number of squadrons they were made mandatory, including Beaufighters in shipping strike wings and Fleet aircraft operating under the Command, such as Avengers and Swordfish on night strike operations off the Channel coast.

The white for coastal continued in post-war years, and Lancasters—black all-over, except for upper surfaces in dark brown and green in Bomber Command—went post-war into Coastal Command in the maritime reconnaissance role re-painted white. The sole exceptions were the Lockheed Neptunes that served in Coastal Command from 1952 to 1956, which retained their US Navy midnight blue overall.

In view of all the experience gained in camouflage during the war, it is strange that the US Navy decided that their post-war patrol aircraft should have dark blue undersurfaces and grey upper surfaces. Nothing could be more conspicuous to surface vessels, but not until October 1964 did the US Navy change this to white upper surfaces and gull grey undersurfaces—a shade similar to present-day RAF Shackletons.

RAF coastal aircraft colours in recent years reflect the functional nature of colour schemes. For example, Shackletons changed to an overall glossy dark sea grey. This grey is the shade now deemed to give the best camouflage, but in a matt application. However, a glossy finish reduces drag—and

A Dutch Navy Fokker C.XI-W with rippled camouflage for operation in the Netherlands East Indies in 1936

This 1942 photograph helped to confirm the superiority of light colours for CC aircraft. The top Wellington is in standard Bomber Command markings, plus underwing roundels; the lower is white

Standardisation in the 'thirties: Fairey IIIFs of No 202 (Flying-boat) Squadron, which operated floatplanes for a period, over Malta

A Shackleton in the 'sixties, in glossy dark-sea-grey, with white-topped fuselage and red serial and squadron markings outlined in white.

since the Shackleton has about 5,000 square feet of surface area this is no mean consideration.

The fuselage tops, and areas over the wing fuel tanks, are painted white. This too, reduces the effectiveness of the camouflage when viewed from above, but white, being light reflective, does not absorb heat—another important consideration where large areas are involved.

On the other hand, Coastal Command operate the most vividly painted aircraft in the RAF, in the shape of the Whirlwind helicopters used for search and rescue, with their high-visibility overall brilliant yellow. Thus, Coastal colours are sometimes far removed from sea shades; but whatever the colouring there is usually a reason for the particular shade and perhaps a story behind how it came to be adopted.

A P-3 Orion in its original US Navy "blue whale" finish. How conspicuous this was to submarine crews is evident by comparison with the white-painted rear fuselage

American shades took over on RAF coastal aircraft in 1952-56, when Neptunes in service retained their US Navy midnight blue

With the US Sixth Fleet

photographs by Brian M. Service

The 75,900-ton carrier USS *Forrestal* was the first of eight ships of her class. The photographs on this and the next three pages were taken during a tour of duty with the US Sixth Fleet in the Mediterranean

Carrier Air Wing 17, assigned to the *Forrestal*, consisted of VF-11 (the Red Rippers) and VF-74 (Bedevillers), each with 12 F-4B Phantom IIs; VA-152 (Mavericks) with 14 A-4B Skyhawks; VA-15 (Valions) and VA-34 (Blue Blasters), each with 14 A-4C Skyhawks; RVAH-12 (Speartips) with five RA-5C Vigilantes; a detachment from VAH-10 (Vikings) with three KA-3B Skywarriors; and four E-2A Hawkeyes of VAW-123

Left: A Phantom of VF-74 follows a Skyhawk of VA-15 into the air. Launch rate is one aircraft every 17 seconds, from two catapults

Below: The Phantom II can carry up to 16,000lb of stores under wings and fuselage. This aircraft has underwing Sidewinder air-to-air missiles and drop tanks

First aircraft to be launched when these photographs were taken was a Douglas KA-3B Skywarrior tanker of VAH-10 laden with 10,000 gallons of JP5 fuel. Its primary task is to top up the tanks of strike aircraft on long-range missions. In addition, one A-4 out of twelve is usually fitted out as a buddy tanker

This COD (Carrier On-board Delivery) C-2A Greyhound of VR-24 flew out a spare engine from the US Navy Air Facility at Naples

Mightiest aircraft carried on board ships of the US Navy, the 30-ton RA-5C Vigilante reconnaissance attack, plane is catapulted off at 125 knots, afterburners blazing. With a top speed above Mach 2, the RA-5C carries cameras, radar and infrared sensors for reconnaissance, intelligence and countermeasures duties

The saucer-shaped radar scanner of this E-2A Hawkeye of VAW-123 has a diameter of 24ft. The radome rotates in flight at 6rpm. For stowage on board ship, the radome lowers 2ft and the wings fold as shown. Hawkeyes provide early warning of approaching enemy ships and aircraft and can also control up to 24 aircraft as forward command posts to direct air strikes

One of the *Forrestal*'s three Kaman UH-2A Seasprite helicopters which are flown by crews of Squadron HC-2 on plane-guard duties. A piston-engined C-1A Trader is carried to ferry mail and supplies between the carrier and the shore or other ships

The Skyhawk is so small that it does not need folding wings for transport by lift to the below-deck hangars. However, it can carry up to 5,000lb of external stores

Two Vigilantes, six Phantoms and 15 Skyhawks parked aft on the *Forrestal* before an exercise

Skyhawk coming in to land on the deck of the *Forrestal*, with its arrester hook, wheels and flaps down. Landing speed is 120-135 knots

F-4B Phantom II of VF-74 at the end of the catapult run on take-off. A mixed armament of infra-red Sidewinder and radar-homing Sparrow III missiles is normally carried

They Followed the *Flyer*

by Maurice Allward

Wright *Flyer* No 3 in flight, October 7th 1905. At this time the pilot still lay prone on the wing, as in the first historic hops of *Flyer* No 1

ONE OF THE best remembered of all aircraft is the Wright *Flyer*, the world's first successful powered aeroplane. Much of this fame is due to the one and only photograph taken at the moment of its first historic take-off in December 17th, 1903. This photograph has, not unnaturally, been used in almost every book covering the development of flight, and will continue to be so used. Unfortunately, such "over-exposure", while helping to immortalize the *Flyer*, has tended to detract attention from the long line of interesting aircraft which Wilbur and Orville Wright evolved from this first machine.

The *Flyer* itself was a direct development of three gliders. The first of these, a sturdy biplane with a simple front elevator and wing warping for lateral control, was made in the year 1900. It was taken to Kitty Hawk, where a number of successful glides was made, but was mostly flown and tested as a kite, with the controls operated by cords from the ground.

As this glider did not have the performance that had been expected, a second model was constructed in 1901. This had the wing area enlarged to 290sq ft, and the camber of the wings was increased. The deep curvature caused trouble and disappointment, and in the following year Glider No 3 was built. On this machine, the wing area was increased further to 305sq ft, and twin vertical fins were fitted in addition to the standard forward elevator. After various modifications some successful glides were made, but the aircraft tended to get out of control when the wings were warped.

Gradually, the Wrights realised what was happening—when a wing was warped to increase the lift, the drag, as well as the lift, increased. So the brothers changed the twin fins for a single movable runner and linked its control wires to those for warping the wing. Thus, when a wing was warped to raise it, the rudder was automatically put over to counteract the drag resistance and take the machine round in a smooth banked turn. This brilliant idea gave the Wrights the key to success; together with the forward elevator it enabled them to perform all the basic flying manoeuvres. With the improved glider they made nearly a thousand successful flights. It now remained only to add power.

This they did in 1903. They built a new airframe, on the lines of the 1902 glider, but with the wing area increased to 510sq ft and a twin elevator in front as well as twin rudders at the rear. This was the historic *Flyer* which on December 17th, 1903, made the first sustained, powered and controlled flight in the history of the world. The *Flyer* made three more flights that day. After the fourth, a gust of wind overturned it, the resulting damage being too serious to be repaired on the spot.

Returning to their home in Dayton, the brothers started designing and building a second machine, with a new engine. Known as No 2 *Flyer*, it was basically similar to No 1, the wing area being slightly increased, the span increased from 40 to 41ft and the camber decreased. In addition, and most important, it had a more efficient 16hp engine and different gearing to the propellers.

The new machine was completed in May 1904 and taken to a field near Simms Station, some eight miles east of Dayton. Named the Huffman Prairie, this field was used for the first tests of the No 2 *Flyer* on May 26th. During the following months more than 100 flights were achieved, including the first complete circuit. To make the take-offs less dependent upon the weather, and to offset the small size of the field, the brothers developed an ingenious falling weight catapult launching device, which was first used on September 7th. The longest flight of the season was made on November 9th, when Wilbur flew for over five minutes, covering nearly three miles.

The year 1905 saw the appearance of a third, still further improved machine, *Flyer* No 3. The wing area was slightly reduced and the camber put back to 1 in 20, as on the original *Flyer*. The wings were rigged horizontally flat, without the characteristic anhedral droop of the earlier aircraft. The front biplane elevator was enlarged and moved forward, and the twin rear rudders were also increased in size and placed further back. New propellers were fitted, but the engine from the 1904 machine was retained.

Again using the Huffman Prairie as base, the new machine made its first flight on June 23rd, and trials continued until the middle of October.

One of the problems encountered during 1904 on the No 2 *Flyer* had been the tendency for the aircraft not to respond to warping when making a tight turn, so that it stalled and went out of control. This phenomenon the Wrights could not understand or solve at the time. While seeking a solution on the No 3 *Flyer*, the brothers separated the wing

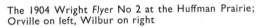

Type A two-seat *Flyer* in front of hangar at Hunaudieres race-course, near Le Mans, in August 1908

The 1904 Wright *Flyer* No 2 at the Huffman Prairie; Orville on left, Wilbur on right

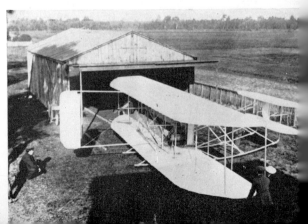

Orville and Lt Selfridge prepare for take-off on the flight that ended so disastrously, September 17th, 1908

Wright *Baby Grand* racer of 1910, with wheeled landing gear in place of the skids used on earlier types

The Wrights' 1911 glider in flight at Kitty Hawk, North Carolina

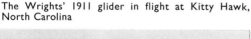

The Model D was similar in configuration to the 1910 racers, but larger

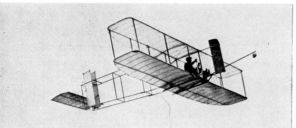

warping and rudder control wires, so that they could be operated either independently or together. At this stage the Wrights realised the nature of the stalled condition on a tight turn—and that it could be cured by putting the nose down and increasing speed to maintain controllability. The separation of the wing and rudder controls also permitted the re-establishment of lateral equilibrium in gusts, without the automatic turning movement induced by the rear rudder when the controls were interconnected.

This freeing of the controls enabled the brothers to enhance still further their mastery of flying. Altogether, a total of 49 flights were made during the 1905 season, the longest being one of 38 minutes during which 24 miles were covered. The flight ended only when the fuel was exhausted.

A strange aspect of these two years of intensive trials, in *Flyers* No 2 and 3, was the almost complete absence of press reports on what were obviously technical achievements of the first order. Octave Chanute, a

pioneer glider pilot, who had witnessed a flight at the Huffman Prairie on October 15th, 1904, later wrote to Wilbur commenting: "It is a marvel to me that the newspapers haven't spotted you".

The Wrights did, in fact, invite witnesses and pressmen to attend the early 1904 trials. Some fifty people, including twelve journalists turned up at the field. Unfortunately, by the time the brothers had completed their preparations, the wind had dropped to a gentle breeze, not sufficient for taking off from the short track then used. In order not to disappoint the people present, some of whom had come a long way, an attempt was made. As if the lack of a reasonable wind was not bad enough, the engine now started to misbehave. The machine ran along the short track . . . and off the end without rising into the air at all. Disappointed, the onlookers departed. A number returned the next day, including several reporters, but were again disappointed. The engine continued to give trouble, and after a short glide the machine slithered to rest on the ground. The Wrights decided to abandon further attempts until the engine could be put right. The reporters, dis-illusioned, left the field and lost further interest in the machine.

Their disinterest continued even when reliable reports of successful flights of several minutes duration trickled in over the next two years. This seems incredible to us in this age of instant news; but it is under-standable if one recalls the condition at the time. Even though the sight of *Flyers* in the air was commonplace to the local farmers, there were still reputable authorities who asserted that mechanical flight was impos-sible, and many reporters had experienced abortive visits to witness the flights of crackpots whose strange contraptions did not, and could not, ever fly.

Wilbur explained that when the reporters later heard that "we were making flights of several minutes duration, knowing that longer flights had been made with airships, and not knowing any essential difference between airships and flying machines, they were but little interested". One local Ohio newspaper editor admitted, however; "I guess we were just plain dumb".

Historians have speculated that but for this disinterest aviation development might have taken a quite different course. As it was, with the Wrights' powered experience unavailable to other would-be aviators, development elsewhere moved along inde-pendent but slower lines. Information on the Wrights' gliders was more readily available, and led to the development of half-a-dozen close copies in Europe.

During the 1905 season, the brothers tried to interest the United States Government in their progress on more than one occasion, without success. Looking overseas for recog-nition, they fared as badly at the hands of the British Government. Frustrated, and confident that no-one else would be able to

fly for years, they stopped all flying at the end of 1905, to prevent spying and to give them time to open negotiations with anyone who seemed interested. Thus, for more than two and a half years, from October 1905 to May 1908, the two pioneers remained grounded and permitted no-one to see their aircraft.

During 1906 the brothers built a number of new engines but no aeroplanes. In 1907 they built more engines and a number of new aeroplanes, but still did not fly at all.

In May, Wilbur took one of these 1907-type *Flyers* to Europe, to negotiate with European companies. Orville followed later, and the two brothers visited England, France and Germany, with no results. Wilbur returned to America, to start a new programme of construction. Orville returned later with, at last, hopes of selling the design to a European company. The precious *Flyer* was left in its crate at Le Havre.

Early in 1908, after three lost years, things happened rapidly. In February, the US War Department at last agreed to witness an official test, and in March an agreement was reached for Wright aircraft to be built in France.

To prepare themselves for the official tests, the brothers took the 1905 *Flyer* (No 3) to Kill Devil Hills for practice. At this stage, a seat was fitted, so that the pilot no longer laid prone on the bottom wing, and a second seat was also installed so that a passenger could be carried. Some twenty-two practice flights were made on the modified aircraft.

Wilbur then returned to France, to uncrate the *Flyer* at Le Havre and prepare it for flight. The trails were to be held at a race-course at Hunaudieres, about five miles south of Le Mans. Preparations were com-plete by August, and on the 8th of that month the first public demonstration was made. In the succeeding days eight further flights were made, followed by another 100 from nearby Camp d'Auvours military ground.

Last product of the Wright company was the Model L single-seat military scout of 1915

Powered by a 60hp Wright engine, the Model C-H of 1913 was intended as a commercial hydroplane

The results were electrifying. There was no limit to the amazement and no limit to the praise. It was not just that the *Flyer* broke every existing record for aeroplanes, but that here was an aeroplane that could climb, bank, turn, execute smooth circuits and figures of eight and even demonstrate glides with the engine shut off. The aircraft and flying technique were far in advance of anything else in Europe at that time.

Meanwhile, Orville was achieving similar success with his US Army trials at Fort Myer. Of three aircraft ordered from different designers, his *Flyer* was the only one actually delivered and in it he made ten flights, including four of more than an hour's duration. Then tragedy struck. While making his third flight with a passenger, Lt T. E. Selfridge, a propeller failed, damaging the rudder, at a height of 75ft and the machine crashed, killing Lt Selfridge and injuring Orville. This was the first fatal accident to a powered aeroplane.

In spite of this mishap, at least six or seven similar aircraft were built between 1907 and 1909 and were put to good use in extending the Wright's techniques and achievements. Known as the Type A, they were efficient two-seaters, with an ample safety margin in view of the high standard of design of the airframe and propellers. The first was built for Wilbur's demonstration in France in 1908 and the second for Orville's flights at Fort Myer, ending in the crash on September 17th, 1908.

The Wright A machines built in France and England in 1909-10 were similar to the 1907-09 types built by the Wrights in America. Shorts built the six machines constructed in England.

Although these aircraft were capable of taking off unaided, they were normally launched along a rail with the assistance of the familiar derrick and falling weight device. It is not clear why the Wrights retained their skid undercarriages for so long as, even if wheels were not essential for landing and taking-off, they obviously facilitated ground handling. As it was, the first experimental fitting of wheels to the standard skid undercarriage was not made until 1909, when a *Flyer* was so modified in France. The first Wright aircraft to embody a wheeled undercarriage from the start was the little biplane racer of 1910.

Wheels were also fitted to several Type A *Flyers* in 1910, as were a number of other improvements. It was in this year that the Wrights made important changes in their control system. Up to this point all *Flyers* were inherently unstable, in that they had no rear horizontal tail surface. The forward positioning of the elevator was due to the Wrights' belief that this location gave the safest fore and aft control, especially when a large rearward movement of the centre of pressure over the wings caused the machine to nose-dive.

First change to the control system was to fit a fixed tailplane behind the double rudder. Soon afterwards this was converted into an elevator, working in conjunction with the front elevator. Thus modified the *Flyers* were known as the modified standard type. These were followed by the Wright Model B, on which the distinctive front elevator was abandoned. *Jane's All the World's Aircraft* recorded picturesquely at the time that a "headless" Wright had been produced. This model also embodied a wheeled undercarriage. It was one of the earliest American aircraft to be put into production, and got the US Army off the ground at last.

The little racer already mentioned was built in two forms, both without forward elevators. One, with a wing span of 26ft 8in is known as the *Baby Wright*. The other, with an 8-cylinder engine of 60hp and reduced-span wings, was the *Baby Grand*. The latter, built for the Gordon Bennett Race, was to be flown by Brookins, one of the most experienced pilots of the American Aero Club. Although the brothers had given up demonstration flying, Orville, who loved flying, took this machine up on a test flight at Belmont Park and made a circuit at 75mph—about double the speed of their standard machines. In the race itself, Brookins crashed and the race was won by Claude Grahame-White, flying a Blériot at 63mph. At the end of 1911, the brothers returned to Kitty Hawk to try some soaring. They had constructed a new glider, mainly from parts of two old 1905 and 1908 machines, found in a crate. With this glider a number of soaring flights were made, and Orville set a new world record by staying up for 10 minutes. To make rapid changes in the centre of gravity, the brothers used the simple expedient of a rod projecting ahead of the wing, upon which they hung bags of sand of varying weight. Asked on one occasion the purpose of the device, Orville replied mischievously, "It's a special stabilising device". In no time a rumour spread far and wide that a new and secret method of ensuring the stability of an aeroplane was being tried out at Kitty Hawk.

The next major improvement made to the Wright series was the adoption of ailerons. Although ailerons for the lateral control of aircraft had been thought of and patented as far back as 1868, by Matthew Boulton in England, the Wright brothers were unaware of this and had invented and developed wing warping independently. It is such a short step from warping to hinging that one wonders why the Wrights did not quickly develop ailerons; even when ailerons had been perfected by others, they persisted in their preference for warping. In fact, not until 1915 did the first Wright machine (the Model K seaplane) appear with ailerons.

Other Wright aircraft produced during the 1914-18 War years included a two-seat trainer. It was known as a Beatty-Wright, and Alan Curtis, Captain RFC, recalls a trial lesson in one at the Beatty School of Flying at Hendon, in 1914:

"The 'lesson' consisted merely of feeling the controls which, on this particular machine, comprised dual wheels and rudder bars. The pilot was Edouard Baumann, a Swiss. I do not know what modifications justified linking the name of George W. Beatty with the standard Wright biplane of the day, and can only surmise he modified the controls to suit school requirements.

"Take-off was short (despite an engine developing no more than 35hp) and we reached about 900ft according to my logbook —there were no instruments. The aircraft was famous for vertically banked turns and was apparently easy to fly and very manoeuvrable.

"The flight was to consist of two circuits of Hendon aerodrome but the engine was misbehaving on that day and, before the circuits could be completed properly, we had two forced landings, which added considerably to my joy!"

The model L appeared in 1915. Powered by a 70hp six-cylinder Wright engine, this was a single-seat military scout. It was a neat-looking biplane, with ailerons on both mainplanes, and was the first Wright aircraft to have a single tractor propeller. The Model L was also the last aircraft designed by the Wright Company, and brought to an end the long series of types evolved directly from the original *Flyer* which, twelve years previously, had ushered in the aviation age, the full portent of which has yet to unfold.

Stringbag Days

by E. C. Cheesman

AS I DRIVE along the seething Oxford Road towards London these days, I often think of a summer's morning just over 50 years ago when it was a tree-shaded lane and our squadron, in full battle rig, marched the four miles from Denham to Uxbridge.

It was July 1918. The war in France was going as undecidedly as ever, and to add to the troubles that unwanted baby the Royal Air Force had just been born. Having been accepted, to my delight, for service as a "Probationary Flight Officer" in the Royal Naval Air Service, I had a bitter disappointment one morning to receive a badly stencilled sheet of foolscap informing me that this famous service had ceased to exist, and that I could enter if I wished as a "Flight Cadet" in the new Force. Take it or leave it, as it were.

Shortly after we (there were about a hundred victims) buried our pride and joined the first course of the Initial Training Wing at Hastings. There I worked harder than I had ever done before, under the command of that well-known figure, newly arrived from Canada, Brig-Gen Critchley.

I must say that I thoroughly enjoyed those Hastings days. We did exhausting cross-country runs before breakfast and forced marches in between lectures in pier pavilions and public gardens where archery was normally practised. We learned our Morse in an Infants' school at a place called Bo-peep and got seven days confined to billets for omitting to polish the soles of our spare pairs of boots. After this came the School of Technical Aeronautics, near the golf links at Denham, where we were lectured on engines and airframes by Malcolm Campbell, who had built a lean racing car for which we somehow contrived a concrete speedway strip.

Uxbridge School of Gunnery followed—probably one of the first courses held there—and this led to a posting to my first aerodrome, Harling Road, near Thetford in Norfolk. It was a new station, sited alongside the main railway to Norwich. There were some fine concrete hangars and a block of barrack quarters was going up, but I arrived after dark, in the rain and, finding nowhere to report, bedded myself down in an unfinished hut.

There were, at first, more would-be pilots than aircraft. They included South Africans, French Canadians and New Zealanders, and the ground work was done by American fitters and riggers. We even had German prisoners for batmen. The uniform was

Aircraft operated by No 10 Training Defence Squadron included Avro 504s (*left*) and R.E.8s. The photographs illustrating this article were taken by the author in 1918

almost entirely that of the Army and RFC and I felt very conspicuous in my RNAS dark blue, which was all I possessed at the time. The new light blue garb of the RAF had been designed and approved, but it was some time before we saw anyone wearing it as no grants were available and we couldn't afford to buy our own.

After a fortnight I had not even been posted to an instructor and my first flight was quite unexpected. One day Capt Reeves, the Wing Examining Officer, appeared at lunch. I was sitting near enough to hear him say that he was on his way to the Naval Airship Station at Pulham, some 40 miles away. After the meal I sidled up to him and asked if he would let me accompany him in his Avro 504. Very much to my surprise he agreed, and within half an hour I was in the air and interestedly watching a farm cart below.

But I wasn't given much time to look at farm carts, for as soon as we were on an even keel he demanded over the speaking tube how many hours had I flown. I couldn't even answer, as I hadn't got my end fixed up, but there was no need to. With the 1918 equivalent to "it's all yours" he handed me the Avro, and the next half-hour was a bit of a nightmare, to say the least of it. However, the great airship sheds at Pulham soon came into sight and we began to come down, finally landing on the wrong side of a dyke, a mile from the sheds.

Dead silence prevailed, but after a time a truck appeared and took us over to the Naval Mess (dark blue at last!). We had tea and the CO, Commander Little, took us over to the colossal hangars. Three "rigids" were inside—R.23, R.25 and R.26. The keels were 30ft above the floor, the height 120ft and overall length 650ft. As we looked up at the great shadowy bulk of R.23 I remembered Little's words, shrugged as he left the hangar: "These things will be the death of me one day". Alas, they were: he

was killed in the R.38 disaster over Hull.

It was now 16.30 and the light was fading, the clouds were very low and rain had set in. To make matters worse, the engine refused to start, and as none of the naval ratings knew anything about aeroplanes we had to swing the prop ourselves until it finally fired.

We got off the ground and headed in the direction of Harling Road, but at 500ft we were in black cloud and the engine was running badly. After all these years I haven't the haziest idea what instruments there were in old Avro E8194 but I don't believe it even had a workable compass. All I was sure about was that we would not be likely to reach our destination and Capt Reeves was no doubt thinking the same thing. Anyhow, to my great relief, after ten minutes of seeing nothing at all, he called: "What about it? Shall we go on?" This time I had my phones working and although it was my first flight and he the Wing Examiner, I summoned up sufficient courage to venture "Would it be better to go back, Sir?"

The answer was a prompt 180° turn, but at 600ft we could see nothing but a pall of black and there didn't seem to be any breaks. For some more endless minutes things looked pretty bad but all of a sudden a great hole was gashed in the clouds and, framed in it, by a miracle, were the great airship sheds below. We whipped through that hole like an arrow, saw the welcome grass rising up to us and a moment later trundled safely to a standstill. The Navy entertained us suitably that night but admonished us severely for venturing about in such dangerous things as string-and-plywood Avros.

Training in earnest started soon after this, and life on an airfield during the second World War reminded me in many ways of those days in 1918. The aircraft of No 10 Training Defence Squadron consisted of, besides Avros, a few R.E.8's, which were even then obsolete, some slow-flying D.H.6's (they could almost stand still in the air against a strong wind), Sopwith Camels and Pups, Snipes, S.E.5's, and D.H.4's and 9s.

The last arrival I remember was a D.H.9A with American Liberty engine. This landed one afternoon in the hands of a test pilot from Norwich. He had tea in the Mess, thanked us for our hospitality, and departed on the 16.05 train back to Base. We realised after he had gone that none of us had asked him anything about the machine and he had modestly vouchsafed no information. Next day, of course, there was a rush to fly the thing. Flaps and brakes, like Handling Notes, did not exist in those days and the first pilot who tried to land it found that he was overshooting by half a mile and went round again. This happened three times, but at the fourth attempt he just managed to bring it in at the farthest end of the field. After this there wasn't a great deal of interest in the D.H.9A until a rather morbid instructor from Nottingham made a pet of it and gradually came to appreciate its qualities.

This same instructor gave me my first lessons and we flew soberly (at from 40 to 80mph) over the Norfolk countryside which often looked eerie in the winter sunsets. Generally the horizon was but faintly defined and it was difficult to fly straight and level. The ground was lost in a purple haze and, owing to excessive rain, whole fields were under water which shone like gold in the mist.

But there was another side to my stolid instructor. Sometimes, without any warning, he would launch into the wildest aerobatics, loops and dives, and I can still hear the scream of the flying wires as he once spun an Avro 504 down almost into a farmyard. I was just praying that we would come out of the manoeuvre when his voice came over the phones in a silly giggle: "Put the wind up the chickens, eh?"

Of other excitements there was the moon-

Refuelling a BHP-engined D.H.4. The pilot sat between the engine and fuel tanks, too far from the observer for easy communication

light night when diners in the Mess were startled to hear the approaching splutter of an aircraft engine. We all rushed out and there, outlined against the sky, hovered the rotund bulk of S.S.E.35, one of H.M's non-rigid airships. The engines stopped and Commdr Little (for so it turned out to be) called down for a landing party. The boys from Pulham were paying us a return visit! We all rushed out and pulled her down to earth by the wispy ropes they threw out and three very cold, muffled, and suspiciously cheerful figures climbed out of the flimsy gondola.

A good party followed and I was among those who went out to speed them homeward, all by this time distinctly the worse for our hospitality. One of the crew laboured at the starting handle of the 80hp Renault engine, while the Skipper fiddled with the controls. Not a spark came until a voice queried anxiously in the darkness: "Got the switch on, Skipper?". "Hic! Nor I have", followed by a roar as the engine came to life. We clumsily eased them up into the sky and I suppose they all got safely home, as we never heard to the contrary.

Fastest British aeroplane in production in 1918, the Martinsyde Buzzard could attain 145mph

For wireless we had a rudimentary form of morse-code-tapping radio. Some of the D.H.4's were fitted with an outside metal drum and a 30-ft length of wireless aerial (which we sometimes forget to wind in before landing). There was also a bombing range nearby at Lakenheath, to which we would repair with our 2-ft long dummy bombs in the racks below the fuselage.

An aerial photography test of those days consisted simply of photographing a "tee" or cross laid out in a neighbouring field; but owing to the shortage of aircraft and the period of only three months' flying training allowed to pilots before they were expected to be ready for operational flying overseas, this test began to hold up the passing out of "fully fledged" airmen until some bright spirit thought of a simpler procedure. The camera was hauled up to the top of one of the 60-ft hangars and from then on we simply took pictures of a miniature tee in the grass at the foot of the building!

Looking back, the training of those days was pretty rough and ready. As far as I can remember no-one made any effort to check over an aircraft before attempting to fly. If a machine was left standing on the apron unattended it was presumed to be serviceable; no such thing as cockpit drill existed and one wonders that there were not more accidents than actually occurred. In many cases, of course, in a heavy landing at 40mph or even less, the old "stringbag" would fall apart round the pilot, leaving him unhurt.

I still have today the log book of the daily Officers of the Watch covering the last months of the 1918 War, and it is astonishing to read how these frail aircraft were daily falling out of the sky with so little harm to those who flew them. The following entries cover just 48 *hours*!

"Sept 29, Flt/Cdt Casserley crashed Avro E2014 while landing. Prop, u/c front skid, leading-edge damaged. 6.45 am.

2.45 pm F/C Cooper whilst landing smashed shock-absorber.

3.50 pm Lt Eastwood and Sgt Knox smashed u/c, Avro C5906.

6.00 pm F/C Damant force landed in field SW of drome. Engine failure. Nothing damaged, pilot unhurt.

"Sept 30, Lt Hollis, with Cpl Lundberg as passenger, crashed in field SW of drome on Avro D7148 at 7.20 am. Smashed prop, undercarriage, front skid, fuselage. Nobody hurt.

8.30 am Capt Kingford force-landed 2 miles S of airfield. Smashed u/c and main plane. No-one hurt.

3.45 pm Lt Richmond crashed Avro E4167. Prop gone, front skid, undercarriage, leading-edge of lower plane. Nobody hurt".

Then came Sunday November 10th with the brief note in the log Book; "The Allies have handed to Germany the terms of Armistice and given them until 11.00 tomorrow to sign or refuse. Flying washed out in afternoon on account of high wind and rain.—D. G. Cameron, F/C".

Next day came the last wartime entry:

"Word received of the signing of the Armistice. The end of the War, practically. God Save the King. Flying washed out. During the remainder of my tour of duty there was nothing to report—T. Barrett, F/C".

I often think of "T. Barrett" and his historical entry. Good luck to him wherever he is. I took over from him as Officer of the Watch from 6 pm on Armistice Night, but my notes record only the arrival of "five Avros and a Sopwith Pup" the next morning. They landed at an almost deserted aerodrome, nearly everyone having escaped to Norwich to celebrate the first day of peace.

Conquest of the Pacific

by Roth Jones

THE PACIFIC, largest ocean on Earth, covers 69 million square miles. Since the days of the early mariners it has captured men's imagination and been a challenge to them all. Despite its tranquility, its palm-fringed atolls and mysterious islands, there has always been a danger lurking beyond the horizon.

If we study the Pacific Ocean on a map it is a huge area of water stretching from the Aleutian Archipelago in the north to Cape Horn in the South, from the Ryukus in the west to Valparaiso in the east.

The mariners conquered it 200 years ago, but no airmen bridged it until 1928, for this ocean demanded the longest non-stop flights of them all, calling for aircraft with a much longer range than the normal types then flying. Additional to this, the flight required a skill in navigation and radio operating almost unknown at the time.

Honour for being first to fly the Pacific Ocean went to two gallant Australians, Sir Charles Kingsford Smith and Charles Ulm, assisted by two little-known Americans, Harry W. Lyon (navigator) and James W. Warner (radio operator). Their flight, in June 1928, from San Francisco to Brisbane,

Australia, in a three-engined Fokker monoplane was one of the most remarkable of all the pioneering flights of the 1920's, for there were probably more unknown factors facing them than on most of the other trans-ocean routes. It was not a hair-brained stunt; it was, on the contrary, a well conceived and carefully calculated trip which, though full of incidents, went according to plan.

The worst hazard that faced the four airmen was the weather, as success depended on the ability of their aeroplane, appropriately named *The Southern Cross*, to withstand the buffeting through the tropics and hours upon hours of blind flying "on instruments".

The 7,400-mile flight was completed in three stages—San Francisco to Hawaii (2,408 statute miles); Hawaii to Suva (3,144 statute miles) and the final leg to Brisbane (1,780 statute miles). In making it, the four airmen were so much ahead of their time that it was another nine years before commercial aircraft began their initial proving flights and 12 years before a regular scheduled service was introduced.

The story of the flight of *The Southern Cross* has been told many times, on film, in radio

---

Kingsford Smith's "old bus", *The Southern Cross*, was an early American-built model F.VIIB-3M that had been taken over, minus engines, from Sir Hubert Wilkins, after it had crashed on ice. Engines were three 300hp Wright Whirlwinds

serials, picture books and text books, but perhaps the most personalised story appeared in *The National Geographic Magazine* of October 1928, only three months after the completion of the flight. The two Australians, the flight very fresh in their memory, told their story to a world waiting to hear first-hand how this crew achieved their objective and how they felt on reaching Australia.

These are but a few widely-separated extracts from their own graphic story:

"We had been the first to cross the Pacific and had achieved the ambition of our lives; we flew across a sweep of 4,900 miles of ocean over which the steady drone of an aeroplane had never been heard; we did not lightly regard the chances against us; there was the lack of confidence in us publicly expressed by certain Australian airmen; we carried steel saws to cut off the outboard motors and steel fuselage and turn the wing into a raft; we had to scribble pencilled notes to one another; our hopes of finding Suva had faded a little; the battering by wind and rain gave us no respite as we could not get out of it; at 9,000 feet it was bitterly cold and we were wet by rain and miserable; for a time it took the combined strength of Smith and Ulm at the wheel to hold *The Southern Cross* level . . ."

There are few memorials to this great flight today. A small plaque near the cricket oval at Suva, Fiji, where the aircraft touched down, notes the landing. By far the most impressive monument is the aircraft itself. Today *The Southern Cross* rests in a small, neat museum at Brisbane airport where it arrived from Suva. It is in a remarkable state of preservation and thousands of people inspect it annually. Most of Brisbane's schoolmasters ensure that their students make at least one inspection during primary training and another during their secondary education. When it changed to the present decimal currency, the Australian Govern-

ment illustrated its $20 bill with portraits of two great Australian aviation pioneers— Kingsford Smith and Lawrence Hargrave. Many Australian postage stamps, too, have recalled the flight.

After the Pacific had been conquered, the civil airlines thought seriously about introducing a service, but to them the route was still fraught with problems due to its isolation, long stretches of ocean and few bases for flying-boats or landplanes. Pan American was the pioneer of Pacific civil aviation and this airline's story is in itself almost as thrilling as the *Southern Cross* flight and is typical of the gusto with which the Americans tackled international flying in the 1930's.

As early as 1932 Captain Edwin C. Musick, veteran Pan Am skipper, was planning his airline's first civil flights across the Pacific. The initial surveys took three years and were aimed at linking California with Manila and Asia. The islands of Midway, Wake and Guam had to be staffed and equipped to handle the Clippers then being built for this route. The great days of Pan Am's China Clippers began in 1935, as Captain Musick made a series of flights culminating in a six-day air mail flight to Manila, followed by the first passenger flight one year later. The aircraft used were Sikorsky S-42's, which were modified continually for greater loads and range.

By March 1937 Captain Musick had surveyed the route to Auckland in his then-famous *Samoan Clipper*; but on his second flight nine months later he and his crew lost their lives when the flying-boat crashed whilst returning to American Samoa with an oil leak.

Two years were to elapse before the South Pacific service was re-established. Due to World War II, civil flying across the Pacific was dormant by then, but the route these airmen had pioneered was used by officers and crews of the Royal Australian Air Force Catalina flying-boat squadrons when ferrying

The Southern Cross with one of the commercial aircraft for which it blazed a trail across the Pacific—a DC-6 of BCPA

Sir Charles Kingsford Smith with Arctic explorer Sir Hubert Wilkins

their aircraft from San Diego, California, to Sydney, Australia. In all, 168 Catalinas were ferried over this route by Australians. In 1946, when the war clouds had disappeared, civil aircraft again began flying the South Pacific route—first Pan American then Canadian Pacific.

Today six international carriers fly the route. In Sydney alone, a jet is arriving from or leaving for the USA every few hours without any fuss or bother. It is a procedure little different from that at any European airport. Yet none of this would have been possible if a few brave men, like Sir Charles Kingsford-Smith and his colleagues, had not been willing to risk all to prove that such a flight could become a reality.

A Sikorsky S-42 of Pan American—the airliner that made trans-Pacific passenger flying practicable

Forgotten Flying Machines

by Philip Jarrett

DESPITE the many books and articles that have been published on the early days of aviation, a number of aircraft are consistently overlooked and so remain relatively unknown or, in some cases completely unidentified. One such was first illustrated in *Aircraft Sixty-Eight* two years ago. By a stroke of good fortune, its identity has now been established, and the machine saved from obscurity.

Whilst browsing through an album of news cuttings from the early 1900s that once belonged to R. M. Balston (of whom more later), a piece from the Worthing Gazette of 1904 drew my attention. Depicting the mystery aircraft, and entitled "To Conquer the Air", it described the aerial machine built by George Clout of Durrington. This machine was apparently the result of 13 or 14 years of study, and in the previous year Mr Clout had tried unsuccessfully to interest the War Office in his invention. The paper commented, "With the customary reticence of the home Government Departments they stood aloof".

Because of this, the inventor was unable to continue his experiments, and this may well explain why his patent application, No 13,075 of June 11th, 1903, was abandoned.

Although the craft was large enough to lift a man, this was not advised, as "the materials used in its construction are not those best adapted for the purpose". The overall weight was 350lb, but it was believed

This unidentified ornithopter is thought to be a product of the 1860-80 era. Although obviously impractical, the flying sunhouse was driven by "man weight and muscular power" conveyed through push rods and a rocking beam; the upstroke was spring-assisted. The load on the wing pivot point was stated to be 140lb, and this weight was claimed to lift "one inch (less or more)". A rudimentary rudder can be seen attached to the fore structure of the carriage, and the Victorian gentleman has supplied himself with enough handles and levers to keep him occupied had there been any chance of leaving *terra firma*

Carl Zenker, of Bremen, Germany, designed this "steerable airship", as he called it, to obtain vertical lift from eight horizontal propellers. Its haphazard construction from bamboo rods and cloth took from August 1873 till March 1900, and cost about 40,000 Marks! Herr Zenker stated proudly that the craft, "Requires no balloon to become airborne, and operates at a nominal 6hp". Weighing 660lb and driven by "fluid air and compressed air", a speed of 1km in 2mins (about 18.5mph) was estimated. Horizontal flight was to be achieved by means of twin propellers, and the craft was steered by a single rudder. One could own a Zenker contraption for a mere 10,000 Marks, from four to six months after receipt of the order.

that this could be reduced by half, using lighter metals. Control was maintained in flight by altering the balance of the car and regulating the angle of the wings. A steering tail was fitted at the rear. But perhaps the most interesting device was that which enabled the frame of the car to be enveloped in descent, "thus converting it into a parachute of some 13ft diameter", as a precaution against "derangement of the mechanism".

Clout's flying machine (page 45, *Aircraft Sixty-Eight*)

Of all the basic types of mechanical flying machines yet conceived, the ornithopter, or flapping-wing aircraft, has taken the longest to develop. Only in recent years has it come within the realms of possibility, and a successful ornithopter has yet to be flown

Two machines of this type, constructed jointly by E. P. Frost and F. W. H. Hutchinson, are shown. Frost's first and largest ornithopter was his 650lb quadruplane of 1877. Although he had hoped to fit a 25hp engine, he had to make do with a 5hp steam engine, which was also required to drive the wheels in order to gain flying speed! Costing £1,000 and containing £60-worth of silk in its feathers, the aircraft was built of red willow reinforced with cane, bound and glued somewhat after the fashion of a fishing rod

The second machine, built in 1905, was intended solely as a test model to obtain performance data. It is shown on its temporary carriage. Except for the wings, it was constructed by "Messrs. Pye, scientific makers of Cambridge" and the Cambridge Autocar Co. Power came from a 3-3½hp petrol engine, operating the wings at a rate of 100 flaps per minute. The craft weighed 232lb and, despite "certain crudities in the motive portion", jumped two feet at each wing beat! The span was 16-20ft.

Both machines are preserved, the first by the Shuttleworth Trust and the second by the Science Museum

Much less is known of the next two machines, which were built by R. M. Balston. The smaller machine is a model built by Balston in 1907 to compete for the £250 prize offered by Lord Northcliffe of the Daily Mail for model aeroplanes capable of mechanical flight. Entries totalled 200, and the tests were conducted on the slopes of the Alexandra Palace in north London. The models were later exhibited in the Agricultural Hall, Islington. When the results were announced in April, the first prize was withheld. The winner of the top prize was a young man named A. V. Roe, soon to become famous in the aeronautical world

The second of Mr Balston's ornithopters remains something of a mystery. This photograph appears to be the only one in existence, and the aircraft is apparently a full-sized man-carrying version of his model. A figure may just be made out, in the fuselage, with the trunk of a tree as background. Its reputed date is 1907-8, but regretfully, further details are lacking

Trans-atlantic Mail by Catapult

by
Helmut Wasa Rodig

"AND THIS, gentlemen, is your point of no return"—these were the words spoken by the catapult control officer on the MS *Schwabenland*, one of Deutsche Lufthansa's three catapult carriers, which accompanied the final and decisive actuation of the catapult release lever. The HA 139 seaplane on the starting sled was immediately accelerated by compressed air through the short catapult travel to the required take-off speed. With a total take-off weight of 17·5 tons, the aircraft had to contend with an acceleration of up to 5g.

There was no alternative to this technique for the series of flights between Europe and North America that had been initiated by Deutsche Lufthansa in conjunction with Hamburger Flugzeugbau GmbH. The HA 139, designed for long-range overseas routes, was powered by four Jumo 205 C diesel engines of 600hp each. With two aircraft of this type, the *Nordmeer* and *Nordwind*, Lufthansa tried to obtain a mail service contract for the North Atlantic route. They were unsuccessful only because the US Federal authorities, though they had nothing

comparable to offer in return at the time, stuck to the principle of reciprocity.

Yet, in retrospect, it is clear that twelve tightly scheduled flights made by the HA 139's between Horta in the Azores and New York marked the beginning of air traffic across the North Atlantic.

On September 24th, 1937, at 0610 hours, GMT, 0410 hours local time, its catapult deck directed into the wind, the MS *Schwabenland* was steaming through the rough seas off the Azores. The night was dark, the low clouds almost melting into the lashed-up ocean. The HA 139 D-AJEY *Nordwind* was on the catapult sled. Its crew consisted of Walter Diele, a Lufthansa captain, and myself as chief pilot of HFB at the controls, a radio operator and flight engineer, both experienced long-range aviators.

The final checks were being made. A red light on the dimly illuminated instrument panel indicated that the compressed air cylinders of the catapult were under pressure and that the safety lock on the sled had been released. After a brief instrument check, the engines were run up to full power and the

The 17½-ton Blohm & Voss HA 139 *Nordmeer* (North Sea) leaving its catapult on the MS Schwabenland

The HA 139 B *Nordstern* (North Star) had its span increased by 8ft to 96ft 9¼in, new fins and rudders and other improvements

The *Nordmeer* on its catapult. The two 139s were fitted later with the redesigned fins and rudders of the *Nordstern* and were redesignated HA 139 A

Nordwind's crew gave the "ready-for-take-off" sign to the catapult control officer. The red light extinguished. The crew members, their bodies pressed firmly back into the seats and their elbows backed up by the armrests to ensure that the controls would remain centred in a neutral position, counted . . . . "21, 22, 23" . . . . and three seconds after the red light had extinguished, there came that heavy, though expected impact.

Accelerating to more than 4g, we were catapulted into the darkness. After coming free of the catapult, there was an initial slight loss of altitude and for a moment the white caps of the waves were visible in the glow of the lights of the *Schwabenland*. Gradually, however, the heavily laden HA 139 climbed and gained speed, and soon the first wisps of cloud swept across the navigation lights.

At this moment we thought of the words of the catapult control officer . . . "and this is your point of no return". In fact, there was no point of return for this flight—darkness, rough sea and maximum take-off weight would certainly frustrate any attempted landing.

A variety of tasks had to be tackled by the individual crew members keeping them fully occupied. According to information relayed by the German Meteorological Office at Hamburg, the crossing of the bad weather zone would take about three hours. Through many years of aviation activities we had learned to trust the forecasts given by Professor Seilkopf, who had only a few reliable data available for his weather charts, but who offset this shortcoming with considerable experience and knowhow. Therefore, nobody on board was surprised when, with the beginning of dawn, the clouds broke up.

Suddenly, through scattered clouds, we saw a light in front of us on the port side. "An island?" our eyes asked as they met. Impossible, our position and course indicated only the water of the Atlantic ahead of us. A few seconds later we were flying over a brightly illuminated passenger steamer. Her red and white funnels identified her as a liner belonging to the Hamburg Süd shipping company.

Captain Diele and I had a lot of work to do, which we shared in hourly shifts. At an altitude of 1,500ft, flying on compass and turn-and-bank indicator through the clouds, the seaplane was exposed to heavy gusts. Flying over the weather was not possible as our service ceiling was limited. Anyway, our diesel engines did not like flying at high altitudes. But from the start they had been running smoothly and consistently.

Radio communication with the *Schwabenland* was excellent and bearings were taken to confirm our exact course—so far a great circle compound course, confirmed by optical drift measuring. With the clouds breaking up and their lower level rising, the wind freshened, reaching an estimated force 8 and blowing unfortunately from 30 degrees on the port side which reduced our ground speed considerably and would lead, consequently, to an extension of flying time. We decided, as was the approved method in marine aviation at the time, to fly at the lowest possible altitude. By utilising the ground effect we were able to minimize the effect of the headwind.

Flying at a low level can be compared in some ways to the technique used by the largest soaring bird of the oceans, the albatross, and each time it was an unsurpassed experience. The impression of speed was increased by the movement of the oncoming waves, and despite the roar of the four engines we almost imagined that we could hear the waves breaking immediately below our aircraft.

Diele and I remembered our earlier journey over this ocean, when we had been for six months members of the crew of the sailing ship *Deutschland*; but our thoughts of the past were interrupted as the flight engineer noted a drop of oil pressure in engine No 3. After a flying time of about six hours we had covered about half of the North Atlantic route and were now north of the sea route to South America and far south of the North Atlantic route, in a "dead zone". As the range of our radio equipment was limited, we were not able to contact either the *Schwabenland* off the Azores or the *Friesenland* off Long Island by radio, and at that time weather ships did not exist, though there were plans for their establishment.

After throttling down the affected engine we climbed slowly to 5,000ft. At this altitude we would have more time for any decision and would also have a much wider field of view, helping us to spot any ship in the area as early as possible. Also, the wind direction had changed by now and we hoped that the tailwind would become stronger as our altitude increased.

Though the symptoms did not necessarily indicate any major damage, we decided for the time being to shut off engine No 3 to prevent further damage and the risk of fire.

A position check with the aid of a sextant showed that we had drifted about 25 miles to the north. This was corrected by a change of course.

Spirits on board had become low, but improved immediately one of our radio calls was answered by the MS *Friesenland*, cruising off New York. Contact with our point of destination had been established after all. Shortly afterwards we saw for the first time that type of sailing craft with the schooner rigging that is peculiar to the sea area south of Newfoundland. For us, the Newfoundland fishermen provided proof that we were at last approaching the coast of the North American continent.

In accordance with instructions from the flight operational controller on board the *Friesenland*, No 3 engine was restarted and its operation monitored continuously. And then we arrived at the eastern end of Long Island. Flying up the Sound, we saw the skyline of Manhattan rising in front of us.

Suddenly we forgot all our troubles. After circling the city of skyscrapers we touched down adjacent to the *Friesenland*, anchored in Long Island Sound, at 20.45 GMT/15.45 EST.

After a flying time of 14 hours 35 minutes this last flight of the season had ended.

Having changed engine No 3, we started the return flight on September 22nd, 1937. It ended at Travemunde after a total flying time of 28 hours, 43 minutes, and a stopover off the Azores, on the 24th. The distance covered was about 5,000 miles. At Travemünde, the *Nordwind* was taken over by Lufthansa's overhaul and check-out team. There were no more flights of the HA 139 over the North Atlantic.

Until the beginning of World War II, the *Nordmeer*, *Nordwind* and *Nordstern*—the latter being an improved version, designated HA 139 B—flew scheduled mail services on the route to South America. By 1940, the six-engined BV 222 flying-boat, with a take-off weight of 50 tons, had been developed by

HFB for Lufthansa's projected passenger service between Berlin and New York, and offered hitherto unknown comfort for passengers and crew, but the war prevented a regular service on this route.

An even bigger flying-boat, with a take-off weight of approximately 100 tons, was under construction at the same period. The flight test programme of this largest aircraft of its time, the BV 238, was started in March 1944, but it, too, had no opportunity to show its capabilities.

Not until May 29th, 1967 did another aircraft developed by HFB circle the famous New York skyline. It happened when Germany's first jet for commercial operation, the Hansa Jet introduced itself to the United States.

Today, giant jets take off for flights between Europe and the USA in a constant stream, by day and night. Equipped with sophisticated equipment and engines, back-up systems, computers, electronic navigation aids, automatic flight control and weather radar, and tracked over the entire distance by efficient air traffic control organisations, these "express trains" of the air cover thousands of miles in minimum time, flying over the weather.

I wonder if the passengers in these modern airlines, or even their crew, ever remember that, 30 years ago, it needed a lot of faith to cover the same distance in 30 hours of hard work in an extremely narrow and uncomfortable cockpit, without any of the aids which make modern flying almost a matter of routine.

Sometimes we should remember that the aviators of that time—Lindbergh, Köhl, Fitzmaurice, Hünefeld and von Gronau— were the pioneers, whose initial successes provided the inspiration for Lufthansa's flights with the HA 139, the predecessor and initiator of transatlantic air travel.

Current product of the company that built the HA 139s is the little HFB 320 Hansa, with unique swept-forward wings

The Hardest Record of All

by Mano Ziegler

THE FAMOUS contests for the Schneider Trophy were over. Three times in succession, victory had gone to Supermarine racing seaplanes, conceived by the renowned English designer, R. J. Mitchell, and forerunners of the famous Spitfire fighter. The last of them, the S.6B, had also raised the World Speed Record to 407mph. For the Italians, accustomed to winning earlier contests, this was one of their most painful defeats, and they spared no effort to turn the tables again. It took them three years to do so. Then, on October 23rd, 1934, an exceptionally streamlined low-wing racing seaplane, the Macchi-Castoldi MC 72, powered by two 12-cylinder Fiat engines with a total output of 3,100hp, took off for a record attempt, with one of Italy's finest pilots, Francesco Agello, in the cockpit. When he touched down, barely twenty minutes later, he had raised the world record for absolute speed to the well-nigh fantastic figure of 440·7mph.

At first nobody thought of trying to beat this record, because no-one believed it *could* be beaten—certainly not by landplanes. The fastest anyone had flown in a landplane at that time was 305mph, and although this record was put up to 352mph in 1935, by a racing 'plane piloted by Howard Hughes, this speed remained far short of Agello's achievement.

There was no mention of Germany in the context of record flying at that time. When the Me 109 and He 112 fighters came on the scene, however, people began to prick up their ears, and when, in 1937, these two machines were shown at the famous Dübendorf Exhibition, near Zurich, in company with the most up-to-date military aircraft of the period, their potential was clear for all to see.

The Me 109, or Bf 109 as it was still called in those days, was already destined to be the standard fighter 'plane of the *Luftwaffe*. The He 112 was every bit as fast, but the 109

The aircraft whose speed record the He 100 had to beat—Francesco Agello's 440mph Macchi-Castoldi MC 72 seaplane

70

had been completed somewhat earlier; in any case, Heinkels were very busy on other work, because the He 111, intended originally for air transport duties, had become a bomber and was in mass-production.

The rejection of his fighter 'plane was, however, a big disappointment to Ernst Heinkel, who much preferred building fast aeroplanes to building big ones, and had made it his life's work to build machines for high-speed flying. So, instead of taking the defeat of the He 112 lying down, he immediately set to work, with Siegfried Günter, his planning engineer, and Karl Schwärzler, his designer, to build a still faster aircraft—the He 100.

The He 100 was the last word in aerodynamic refinement and the first aircraft in which use was made of the explosive-type riveting process devised by Heinkel for wing construction and later adopted all over the world. No one at the Heinkel works, least of all Ernst Heinkel himself, doubted its success; but while work on the prototype was in full swing, there came news of a record flight by the Messerschmitt test pilot, Dr Wurster, who, in the Bf 109V13, with specially boosted engine, had set up a new world record for landplanes, reaching a speed of 379·38mph on November 11th, 1937.

Ernst Heinkel now urged his team to go all out. On January 22nd, 1938, the He 100 V1 prototype taxied out of the shed and took off for its maiden flight. When, not long afterwards, Chief Pilot Gerhard Nitschke (who had also test-flown the He 111) touched down again, he was full of praise for the flying characteristics of the new type. On Whit Monday of the same year, with Ernst Udet at the controls, it set up a world record over a 100-km circuit, reaching a speed of 394·6mph. Both Heinkel and Daimler-Benz, whose new DB 601 engine had made this record flight possible, had done their job thoroughly. But there was a further task ahead—the most important of

all: to wrest the absolute speed record from the Italians.

From the 1,050hp DB 601A, Daimler-Benz technicians developed the 1,800hp DB 601R racing aircraft engine. Preliminary calculations left no doubt that the He 100, equipped with this engine, would be able to reach a speed of more than 435 mph—not it was true, with a standard airframe, but with the version later to be known as the He 100V7 and V8, with the wing span reduced from the normal 30ft 10¾in to only 25ft.

For the preliminary test flights, Ernst Heinkel chose his youngest test pilot, Hans Dieterle, who was only 23. Strictly speaking, Dieterle had been taken on by Heinkel for only six months, as a sort of trainee. But he proved to be such an excellent and reliable pilot that Heinkel was only too pleased to select him for this difficult task.

By early September 1938, the He 100, specially prepared for the world record, was ready for its big day. The engine had been warmed up, and Gerhard Nitschke climbed into the cockpit in order to carry out a short preliminary flight test before the actual attempt on the record. After a normal take-off, the spectators noticed that one of the landing gear legs seemed to have jammed while being retracted. It remained half in and half out, while Nitschke side-slipped and zoomed in a desperate attempt to lower the landing gear or complete its retraction. He failed to do so.

Several times he flew low over the field, in order to communicate by signs with the men on the ground; then he zoomed up, and suddenly the engine was heard to splutter, after which it gave out altogether. Nitschke could not have brought such a fast 'plane down safely, with one landing gear leg partially extended, without endangering his life. He baled out, but was flung against the tail unit, which tore away the pilot chute of his parachute. Fortunately, the main canopy opened just in time, so that Nitschke, although injured, survived the ordeal. But the He 100 and its valuable engine were dashed to pieces on the ground.

While relieved to know that Gerhard Nitschke's accident had not proved fatal, and that he would be fit again after a few weeks in hospital, everyone was dismayed at the thought that all the tests that had been carried out round the clock for months had ended in failure. Ernst Heinkel himself took some time to recover from the shock, but was not the man to give up in the face of a mishap. He ordered a new "record airframe" to be built and approached the Daimler-Benz people for a new "record engine". Finally, he sent for young Dieterle, who now had to take the place of the injured pilot Nitschke.

Dieterle asked for a short time in which to think the matter over and suggested to his employer that the next attempt on the record should be made at Oranienburg. Heinkel showed no great enthusiasm for this proposal, which involved reconstruction of the entire timing apparatus and the transfer to Oranienburg of the record flight team of about 30 men; but in the end he realized the advantages of such a move. Oranienburg airfield was considerably longer than the Marienehe runway, which was less than 3,000 feet long, to say nothing of the railway embankment at one end and a little river, the Warnow, at the other. Furthermore, the 3-km measured course at Marienehe entailed a wide turn over the Baltic after take-off, with consequent lack of satisfactory reference points for the approach to the course. In a record flight of this kind, the take-off and approach were the very operations in which extreme accuracy was required, and Oranienburg offered far better reference points during these phases of the attempt. So Ernst Heinkel agreed to Dieterle's conditions, and the team moved to Oranienburg.

Within a few days, Dieterle, flying a training machine, had selected a record course between Oranienburg and Neuruppin and had it measured; he then started the actual preparatory flights. By the end of February 1939, he had carried out daily flights over the record course in both directions, and every little scrap of forest, every road and every stream were so firmly fixed in his memory that he knew the course and its surroundings like the back of his hand. At last, in March 1939, everything was in readiness.

Jupp Köhler, chief engineer of the Heinkel works at Rostock-Marienehe, an old friend and employee of Ernst Heinkel and a man of outstanding ability and infectious energy, had been appointed leader of the record team. The machine to be used for the purpose, the He 100V8, had in the meantime been completed, and Hans Dieterle now carried out the final trial flights over the course in this machine but with an ordinary DB 601 engine fitted temporarily.

A further four weeks of thorough preparation followed, for this time pilot and ground crew were determined to do everything possible to ensure that the undertaking would not fail again, perhaps irretrievably. Dieterle insisted, for instance, that the special engine built for the record attempt should not be installed in the He 100V8 airframe right away, but should be tried out first of all on a short test flight in another He 100. Only after this flight, lasting about 7 minutes, had been completed without any cause for misgivings did Jupp Köhler feel confident in giving instructions for the engine to be built into the airframe in which it was to be used on the actual record flight.

On the following morning, the V8 was towed to the take-off point, where Daimler-Benz staff had the DB 601R run up with ordinary sparking plugs, as they did in the case of their racing cars, while Herr Stotz and Herr Hoffmann, two of their licenced engineers, checked every detail of the

Model of SSZ 60, one of 66 Sea Scout Zero non-rigid airships delivered to the RNAS in World War I. For further details of exhibits in the FAA Museum, see the article on pages 79-83.

THE FLEET AIR ARM
MUSEUM

Corsair Mk IV KD431; Seafire Mk XVII SX137

Whirlwind HAS Mk 7; Attacker F Mk I WA743

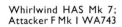

Wyvern TF Mk I VR137; Walrus Mk I amphibian

Swordfish LS326 in flight

RNAS
YEOVILTON
1969

Parent base for all operational RN fighter squadrons, Yeovilton also flies the flag of Rear Admiral C. K. Roberts, Flag Officer, Naval Flying Training, and houses the Fleet Air Arm Museum. As a result, the visitor may see in the air and on the ground reminders of every period of British Naval Aviation history, from an airworthy Swordfish, victor of the attack on Taranto in 1940, to the 1,400mph Phantom that will be the Fleet Air Arm's last fixed-wing combat aircraft   [Rita and James Vanderbeek

Sea Venom of Aircraft Direction Squadron

Sea Vixen folds its wings as it taxies in

Phantoms of 700P Trials Squadron landing

. . and being prepared for the next flight

All take-offs are made with the Phantom's reheat in operation, but its use is limited to the first few hundred feet of climb—out of consideration for the local populace.

SATURN WORKSHOP

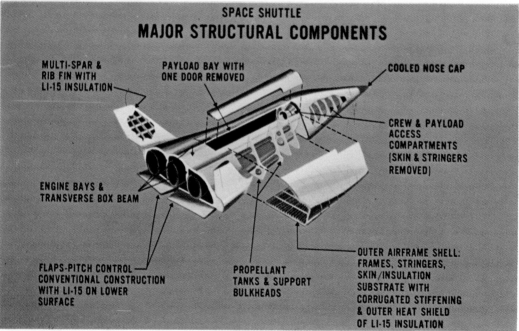

SPACE SHUTTLE
MAJOR STRUCTURAL COMPONENTS

MULTI-SPAR & RIB FIN WITH LI-15 INSULATION

PAYLOAD BAY WITH ONE DOOR REMOVED

COOLED NOSE CAP

CREW & PAYLOAD ACCESS COMPARTMENTS (SKIN & STRINGERS REMOVED)

ENGINE BAYS & TRANSVERSE BOX BEAM

FLAPS-PITCH CONTROL CONVENTIONAL CONSTRUCTION WITH LI-15 ON LOWER SURFACE

PROPELLANT TANKS & SUPPORT BULKHEADS

OUTER AIRFRAME SHELL: FRAMES, STRINGERS, SKIN/INSULATION SUBSTRATE WITH CORRUGATED STIFFENING & OUTER HEAT SHIELD OF LI-15 INSULATION

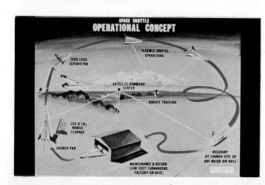

SPACE SHUTTLE TO SATURN

With the first US orbital space stations, like the Saturn workshop, scheduled to be put into space in the early 'seventies, NASA is studying ways of ferrying men and supplies to and from them. One solution is the Space Shuttle lifting-body craft (*centre*), designed to carry seven personnel and 7,260lb of cargo. V-shape wrap-round fuel tanks are jettisoned after take-off, as shown in the diagram (*left*) of a typical launch, orbital rendezvous and return to a conventional airfield

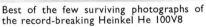
Best of the few surviving photographs of
the record-breaking Heinkel He 100V8

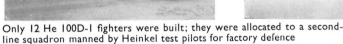
Only 12 He 100D-1 fighters were built; they were allocated to a second-
line squadron manned by Heinkel test pilots for factory defence

engine's operation. Then, while the ordinary sparking plugs were being replaced by the "racing" type, Hans Dieterle climbed into the cockpit and had himself strapped in.

The engine was now started up again, and Dieterle at once opened it up to half to three-quarters of its full power. He now had to check the ignition in all 12 cylinders, and Stotz and Hoffmann did this at the beginning and end of the run-up. The exhaust flame told them that the ignition in each cylinder was normal. They gave Jupp Köhler the agreed sign to let him know that everything was in order. Köhler gave the signal "Cleared for take-off", the chocks were pulled away, and the machine started to move slowly forward—slowly, because the airscrew was specially designed for high speeds and not for high take-off performance.

A potentially dangerous factor was the tendency of the machine to swing off the runway, owing to the high propeller torque caused by the excessively powerful engine— a tendency which increased as the aircraft's speed built up. It was difficult to hold the aircraft on a straight course, particularly since the visibility from the small cockpit windscreen left something to be desired. For this reason, Dieterle had stationed a man at the end of the runway, to wave a large flag and provide him with an accurate reference point during the take-off run.

All went well. The He 100V8 took off after a run of between 2,000 and 2,600 feet, and the landing gear was retracted without any difficulty; but the first time the pilot opened up the throttle, after becoming airborne, the engine began to shake and splutter, mis firing badly. The vibration became so serious that Dieterle pulled the lever back to half-throttle, after which the engine once again ran normally. A return to full-throttle resulted in the same troubles as before, so Dieterle decided to land the aircraft, as it was clear that he could not set up a record under these conditions.

An immediate examination showed that the fuel pumps, which were similar to those on an ordinary DB 601, were inadequate for the record engine, in which the fuel consumption at full throttle, was four times as great. This had not come to light in the ground tests, which had never been carried out for any length of time at full throttle, owing to the delicacy of the racing engine.

The addition of a further geared pump solved the problem of supplying the engine with sufficient fuel, and on the following day Dieterle was able to take off on his second attempt to set up a record. This time the engine ran satisfactorily at full throttle, but after the machine had flown over the measured course once in each direction and had thus been airborne for about 8 minutes, Hans Dieterle was perturbed to find that the oil temperature had risen to about 130°C, which was far above the permissible maximum. Furthermore, the oil pressure had sunk far below the laid-down minimum.

There was nothing for it but to interrupt the flight immediately and return to base with the engine throttled back, as experiments were out of the question at the height laid down for a record flight of this kind, only 70 metres (245 feet) above the ground. It would not have been possible even to escape by parachute at such a height.

The antiquated regulations for record attempts then stipulated four flights over a 3-km course at a height of 70 metres above ground level, and had been laid down during the 1914-18 War, in the days before aeroplanes had superchargers to boost performance at height and when their maximum speed was about 125-185mph. The regulations said nothing about altitude above sea-level, which was a more important factor.

The cause of this second failure was a minor defect in the surface evaporative cooling system, which was quite new, having been designed by Daimler-Benz and Heinkel engineers for the He 100. It was soon remedied, but the question now arose as to whether the short-life racing engine, which had been in the air three times, including the first test flight in another airframe, and had twice been run at full throttle, would survive a fourth flight. Fortunately, the Daimler-Benz engineers had devoted such care to this engine that they were able to approve its use in a fourth attempt. "It's bound to hold out for another 20 or 30 minutes" they told Dieterle, and with this reassuring probability the young pilot, on March 30th, 1939, got into his record aircraft for the third time, while the Daimler-Benz men, again for the third time, removed and replaced the sparking plugs.

Hans Dieterle knew this was the very last chance, as the engine would certainly not stand up to a fifth flight. He already knew that he would succeed in setting a new record if there were no mishaps, as approximate calculations had shown that speeds of over 440mph had been reached during the earlier flights over the course.

He made up his mind not to watch the cockpit instruments too critically and to continue flying even if minor irregularities were apparent.

Once again, the take-off was successful; once again Dieterle pulled the machine up into the record course, then performed a wide turn and opened up to full throttle. He was keeping a careful watch on his instruments, particularly the altimeter and the airspeed indicator. Four times he swung out and then re-entered the record course at 245 feet, with pin-point accuracy, listening carefully to the harsh roar of the powerful engine. Nothing abnormal occurred, and after covering the course four times, in exactly 13 minutes, he set the wheels of the He 100V8 down safely on the runway at Oranienburg.

The team greeted him with rousing cheers, although nobody was yet certain what speed Dieterle had reached. After a first rough estimate, using their own stop-watches, they felt certain it had been 445mph, but Hans Dieterle himself was not greatly concerned with these calculations. He was sure of his success and far too happy to bother about the exact number of mph. But both he and every other member of the team—most of all Ernst Heinkel himself—were nevertheless surprised when, on the following day, March 31st, the exact result came through from the official time-keeping unit: new world absolute speed record, Hans Dieterle in a Heinkel He 100, 463·92mph. Germany and her aircraft industry now had the world's fastest aeroplane and the world's fastest pilot.

Hans Dieterle's flight in the He 100 produced not only the most valuable world record that Germany had so far obtained in the air, but also the one that remained a record for the shortest time. Barely three weeks later, on April 26th, 1939, Fritz Wendel, flying an Me 209 racing 'plane, fitted with a 2,300hp Daimler-Benz DB 601R engine, beat this record by about 5mph in four runs over a higher airfield near Augsburg. The highest speed reached by a piston-engined aircraft, under record conditions, was therefore, and still is, 469·22mph.

Please Don't Eat the Pilot

by Sydney Cooper

THE POSSIBILITY of ending up as the main course at a native feast or the bone ornament around a warrior's neck is not a normal hazard of a pilot's career. However in West Irian—the former Dutch New Guinea—this is only one of a group of unusual conditions which must be faced in an effort now being made to yank the world's second largest island out of the stone age into the twentieth century.

The central part of West Irian is completely isolated from the rest of the world except from the air—there are no roads or any other way to get in and out, but there are dozens of dirt landing strips. For years most of the aircraft using these strips were the Cessnas of various denominations of missionaries; these are now being supplemented by an Indonesian airline called Merpati Nusantara, which had fallen on bad times but is being rehabilitated with the help of the International Civil Aviation Organization, based in Montreal, Canada.

Merpati's problems had been complicated by the financial difficulties which Indonesia has undergone in the past few years; in particular, its small aircraft fleet had been hampered and frequently grounded by lack of spares. ICAO has now supplied Merpati with three short-take-off-and-landing de Havilland Twin Otters, as well as DC-3 spares, and these aircraft are now busily shuttling back and forth between the coast and the Central Highlands, carrying teachers, officials, United Nations experts, and anyone else who can help in the development of the area.

This development would be simpler, of course, if more was known about the territory. Population estimates for West Irian run anywhere from 750,000 to 1,250,000, although a United Nations survey, with precise numerical complacency, lists 479,571 people in isolated clusters along the coasts and 321,000 in the Central Highlands which spear northwestward from Australian New Guinea. That Central Highland figure is, of course, open to doubt; many thousands of people there have never seen an explorer, a missionary or an Indonesian official, let alone

De Havilland DHC-6 Twin Otter of Merpati Nusantara at a typical West Irian "airport". With its ability to take off and land in under 400 yards, this STOL transport is ideal for operation in underdeveloped areas

a census taker. They have also never seen woven cloth or machine-made objects of any kind, except for unusually large birds which occasionally roar their way through the skies.

The Highland natives live in little tribal groups, frequently unable because of language differences to speak to the next-valley tribesmen, with whom their only intercourse is warfare. The United Nations report, abandoning its normal dislike for superlatives, says that the area "operates at a stage of technical primitiveness that is hard to duplicate anywhere in the world". The tribesmen live at a subsistence level dependant mainly upon sweet potato cultivation (132 known varieties) and upon a precarious search for small amounts of protein derived from rodents, birds, grubs and insects. There are no large indigenous animals with the exception of pigs, and these are so scarce that they are eaten only on ceremonial occasions. The cannibalism of West Irian results from the lack of meat, and although the Indonesian government tries hard to discourage this method of obtaining dietary protein, the great unknownness of the country often makes this good intention hard to enforce.

Primitiveness does not show itself only in agriculture. The daily way of life is magic, spirits and taboos. Animals, rocks, trees, grass, virtually all objects animate and inanimate, are thought to be inhabited by spirits, and fear and superstition rule. All this is certainly not conducive to development; no Papuan tribe—at least, no known Papuan tribe— has ever developed metal working or ceramics or the wheel, and only in one spot in the north has woven cloth been found.

The coastal and river areas of West Irian are not nearly so backward. There are a dozen fair-sized towns, mostly with airports and diesel-powered electricity, and even a university. The economy is partially based on money, at least in the towns. Until six

years ago the area was a Dutch colony, and the Dutch subsidized the economy of the coastal regions while the highlands were left mainly to the missionaries. When Indonesia took over in 1963 the Dutch subsidy (which in 1962 had amounted to about £10 million) of course ceased. During the following years, while the Indonesian government did provide some subsidy, Indonesia's own monetary difficulties resulted in the deterioration of many of the existing services in the coastal areas, including sea and air transportation.

In the meantime, the Dutch Government had provided thirty million dollars (£12½ million) as a fund for the development of West Irian under the control of the Secretary General of the United Nations. After consultation with the Indonesian Government, a UN survey mission went to West Irian, and drafted a report for the operation of FUNDWI.

Because of the relatively small amount of money available (equivalent to little more than one year's subsidy during the Dutch regime) it was decided to concentrate on the coastal areas and to try primarily to revitalize the economy by rehabilitating surface transportation and power generation, by improving technical education and by developing such things as small industry, fishing, mining and logging which have the expectation of bringing fairly quick returns, including exports which can be turned into foreign currency. The Central Highlands are not to be completely neglected, and will receive aid in matters such as educational supplies, inland fisheries and the development of forest products for local construction purposes, but the overwhelming bulk of the dollar allocation is for the coastal areas.

The exception to this is civil aviation, to which $4 million was allocated, under the management of ICAO. The programme starts with a group of paved airports around the coast, at Sukarnapura (the old Hollandia),

One of Merpati's DC-3s, for which paved airfields have been provided under the ICAO assistance programme

The Twin Otter, powered by two 579shp Pratt & Whitney PT6A turboprops, carries up to 19 passengers

A typical runway in the Central Highlands, improperly paved and devoid of any facility

Manakurari, Sorong, Mibire and Merauke, all suitable for DC-3's, the keystone being the island of Biak off the northern coast with the only international airport in West Irian. In 1967 Merpati Airlines was offering an irregular service into these points and into parts of the highlands, but its fleet was a mixed bag, including DC-3's, Pilatus Porters, Dornier Do 28's, a Prestwick Twin Pioneer and a Beaver amphibian. The greatest problem was lack of spares—which kept a major portion of the fleet grounded much of the time.

The programme agreed to by ICAO and Indonesia is in three parts: re-equipment of Merpati and provision of aircraft spares and operational advice to the airline; repair and maintenance of the base at Biak, as well as of five coastal aerodromes, together with provision of crash, fire and rescue equipment; and establishment of a radio maintenance base at Biak plus purchase of telecommunications equipment for the subsidiary bases, together with package equipment for two dozen highland strips.

The ICAO project actually began before the rest of the development activities in West Irian, for it was obviously impossible to do very much until the United Nations experts were able to circulate freely. The original ICAO survey team, sent out in mid-1967, consisted of W. C. Krishnan of India, Bertil Hellman of Finland and H. A. Schastok of Germany. Krishnan was subsequently appointed ICAO project manager and arrived at Biak in the autumn of 1967, and the three Twin Otters were flown out to West Irian between September and December, as they came off de Havilland's line; operations began in January 1968.

Eight of Merpati's pilots were sent to Toronto to be checked out on Twin Otters, and about a dozen of the airline's mechanics were also trained there. The technicians and pilots sent to Toronto were all trained initially at the Indonesian Civil Aviation Academy in Tjuruk, set up in 1952 with the help of the ICAO technical assistance programme.

Three other ICAO experts—an airline operations adviser, an electronic field engineer and a mechanical field engineer—are assigned to the project. The operations adviser, W. G. McElrea of Canada, is already at Biak, and the other two will arrive during 1969.

A short note at this point from D. P. (Buck) Taylor, a senior ICAO Technical Assistance Officer who administers the West Irian project, after returning from the area: "In case your geography is weak, Biak is exactly one degree south of the Equator, longitude east 136°, just about as far as one can get from Montreal. Sounds an ideal spot for technical assistance activities, especially when we learned that the one expert (not the one ICAO expert, the only expert!) finds it impossible to spend his local currency. The only things produced on the island are bananas and coconuts. Life is not, however, as rosy as it appears. Snags? No newspapers, books, radio or television; only one plane a week, which comes from Djakarta, nearly two thousand miles away, and practically all food and necessities have to come on this plane. A compensation to quote from the UN Post Report—'Swimming is excellent at all coastal points although many beaches are limited because of coral reefs! Other hazards include small black flies that can inflict very severe bites and, in deep water, sharks! Particular care must be taken in rivers to avoid crocodiles which abound.' It is not surprising that Krishnan has taken up gardening as his spare-time hobby. This is strenuous, but safer than swimming. It consists of hacking a hole in the coral and then bringing soil from some 25km away to put in the hole. Things are brightening up on the food horizon, however, as the Assistant UN Development Programme Resident Representative in Sukarnapura is opening up a UN Commissary. He is comparatively near to Mr. Krishnan, being only some 300 sea miles away!"

The ICAO portion of FUNDWI is now well established, the United Nations and the other specialized agencies are about to begin. What the ensemble will mean to the natives of West Irian is hard to tell; the thirty million dollars is being spent slowly and carefully, with great emphasis on capital development and the establishment of a monetary economy to replace the existing, mainly barter, economy.

For the highlands, even with the increased mobility of the aeroplane, development will be slower still, as immediate return on investment is not so likely. However, there is one way in which the highlands can aid the coast, if air transport becomes more regular, and that is in providing food. The coastal soils of West Irian are poor indeed, with the result that there is great scarcity and little variety of available foodstuffs. With suitable development, the highlands could become the breadbasket, or yam basket, of West Irian, with freighter aircraft (STOL, of course) providing the transport. And the freighters would be filled in both directions, with a flow of food from the highlands to the coast, and a flow of trade goods to the highlands to pay for it.

Some Lesser-Known Air Museums

by Leslie Hunt

THE MAJORITY of readers interested in aircraft preservation will already know the world's major air museums and collections such as the Smithsonian and Air Force Museums at Washington and the Wright-Patterson Air Force Base, the magnificent Rockcliffe (Ottawa) National Aeronautical Collection, the Shuttleworth Collection at Old Warden, England, with its splendid "flying days" throughout the summer, and the historic aircraft in the Imperial War Museum and Science Museum at Lambeth and Kensington, London. Yet many may not yet know of the growing collections now being assembled in places as far apart as Moscow and Melbourne, Palam (New Delhi) and Mundelein (Illinois), Lucerne and Willow Grove (Pennsylvania).

Following dispersal of part of the Tallmantz Collection, several new museums are forming in the USA, but there is still much to see at Santa Ana as well as at Ed Maloney's famous air museum at Ontario, also in California. But let us take a look at some museums which have grown up in the nineteen-sixties, starting with the increasingly-popular Fleet Air Arm Museum at Yeovilton, Somerset.

On the A.30 road, just east of Ilchester, the Royal Naval Air Station Yeovilton has displayed since 1963 a most interesting, and steadily expanding, group of aircraft associated with the Fleet Air Arm, comprising at the time of writing: Supermarine Attacker F.1 WA473, Grumman Avenger AS4 XB446, Blackburn NA39 Buccaneer S.1 XK488, Chance Vought Corsair IV KD431, Westland WS51 Dragonfly WN493, Fairey Firefly AS5 WB271, Grumman Martlet 1 AL246, Supermarine Seafire FRXVII SX137, Hawker Sea Fury FB11 WJ231 (painted as WE726), Hawker Sea Hawk FGA4 WV856, Percival P.57 Sea Prince WF137, D.H. Sea Vampire LZ551 (first carrier jet), D.H. Sea Venom NF21 XG737, Fairey Swordfish III HS618 (as V6105), Supermarine Walrus I L2301, and Westland Wyvern TF1 VR137.

During 1969 the museum hopes to show the Sopwith Baby seaplane 8214 ex-Nash Collection and also the $\frac{5}{8}$th scale model of the Fokker Dr.1 which members of No 209 Squadron, RAF Seletar, Singapore, built from the Lawrence Parasol monoplane which Flt Lt Ray Lawrence of No 66 Squadron built at Kuching in Sarawak. The Fokker Dr.1 model commemorates the great victory of No 9 (Naval) Squadron (later 209 RAF) when Richthofen "The Red Knight" was shot down. Lt-Cdr Les Cox, RN(Ret), is the enthusiastic curator of this fine collection, open April-September, admission free.

At Yeovilton one may also see the flyable Swordfish LS326 (ex G-AJVH), the Aeronca C.3 G-AEFT of the Buccaneer Flying Group and Tiger Moths NL750/A2123, BB731/

Now in the Museum of Transport at Lucerne, the Dufaux biplane made the first flight over Lake Geneva on April 28th, 1910. The 41-mile, 56-minute flight was the longest over water by that date and marked the real beginning of flying in Switzerland

A2126 and DE373/A2127, one of which, we trust, will eventually find a place in the museum for posterity.

Our next call will be at Lucerne, where Mr. A. Waldis and his staff at the Museum of Transport and Communications, Lidostrasse 5, will be happy to show you the Dufaux biplane of 1910, Bleriot monoplane No 23 of 1913, Messerschmitt Bf 109E J-355, Fieseler Fi 1 56C Storch A-97, the Federal Aircraft Factory N-20 prototype and a Chanute hang glider of 1933. Many other fascinating exhibits are in store in various places, awaiting a new building here and also for the Technorama at Winterthur.

From the snows of Switzerland we hop to New Delhi where, at the Indian Air Force Museum, Palam Air Force Station, the main exhibits of aircraft are housed in a large hangar, with old barracks used for displaying equipment, uniforms, and other items. Opened on April 22nd, 1967, the Museum has 17 aircraft, mainly indoors, but with a Liberator on the apron and one or two of the Spitfires at dispersal bays for realism. The collection comprises: Westland Wapiti IIA K183, Westland Lysander IIIT 1589 (from Canada), DH Tiger Moth HU512, Auster V IN959, Consolidated Liberator HE924, Hawker Hurricane IIB AP832 (RAF serial), Spitfire VIII NH631 (RAF serial), Spitfire XVIII HS986, Spitfire XIX HS964, Fuji Kokuki K.K.MXY7 Ohka, Hawker Tempest II HA623, DH Vampire NF10 ID606, Percival Prentice IV336. Sikorsky S-55 IZ1590. Folland Gnat Mk.1

General view of the Fleet Air Arm Museum at Yeovilton. Wyvern VR137 in foreground, Swordfish HS618 at rear

IE1059 and the remains of Sabre 25248 of the Pakistani Air Force.

On now to Australia where, at the Moorabbin airfield, near Melbourne, the Australian Aircraft Restoration Group, formed in 1962, has grown from a small penniless, homeless, group, to an organisation which now attracts around 50,000 visitors each year to its growing assortment of machines in a fenced area of the airfield. Today it offers a Commonwealth Aircraft Corporation CA-1 Wirraway A20-10, Commonwealth Aircraft Corporation CA-18 Mustang A68-105, Gloster Meteor Mk.7 A77-707, Fairey Firefly TT Mk.6 WD828, DH60G Gipsy Moth VH-UKV, DH Tiger Moth A17-377, DH Vampire FB30 A79-422, and BA Swallow VH-UUM. In addition, there are several fine machines undergoing restoration, including Beaufighter A8-328, Anson VH-FIA, CA-6 Wackett Trainer, Percival Proctor 1 VH-AUC, Bristol Sycamore A914, a Curtiss P-40E Kittyhawk, Miles Messenger 2A VH-AVQ (once G-AJKG) and a CAC Boomerang II A46-249 (the last one built). By the time you read this, other aircraft may well have arrived for the devoted attention of this band of preservationists.

Over to the United States now, to visit the Naval Air Station Willow Grove, Pennsylvania. Along the main Highway Pa309 are displayed several unique machines, including a rare Arado Ar 196A-3 twin-float scout 'plane used on board the German cruiser *Prince Eugen* in World War II. Alongside, regrettably the recent target for vandals, can be seen a Mitsubishi Zeke, mark unknown but apparently a late model, a Kawanishi N1K2 Shiden ("George") with two small bombs under wings, a Nakajima B6N2 Tenzan ("Jill" 12), a Kawanishi N1K1 Kyofu (the last "Rex" on large single float), Messerschmitt Me 262B-1A-U1 Bu 110639 (USAF FE610), a Curtiss P-40E in "Flying Tiger" marks but with no serial, Lockheed TV-1 Shooting Star 33924, Chance Vought

Kawanishi N1K1 Kyofu (1,460hp Kasei 13) floatplane fighter; NAS Willow Grove, Pennsylvania

EKW C-35 reconnaissance/close support aircraft (860hp H.S. 12Ycrs); destined for Lucerne

Lavochkin La-15 (3,500lb RD-500 turbojet). Contemporary with the MiG-15; at Monino

Lavochkin La-11 (1,850hp ASh-82 FNV). Last Lavochkin piston-engined fighter; at Monino

Arado Ar 196A-3 scout seaplane (970hp BMW 132K); at the US Naval Air Station, Willow Grove

Kawanishi N1K2 Shiden fighter (1,990hp Nakajima Homare 21); also at NAS, Willow Grove

Only known surviving Nakajima B6N2 Tenzan torpedo-bomber (1,850hp Kasei 25); Willow Grove

Convair XF2Y-1 Sea Dart experimental hydroski fighter; at NAS, Willow Grove

F7U-3 Cutlass 129642, the Convair XF2Y-1 Sea Dart and North American FJ-1 Fury 3568. Let us hope that the authorities will be able to prevent further senseless damage to these priceless aircraft, even if they have to be taken inside with only limited viewing.

Within easy reach of Chicago there are now several venues for the vintage aircraft enthusiast, with many interesting exhibits at the Museum of Science and Industry, Wm. Ross's small but select collection and, for those with time, Chanute AFB at Rantoul.

Yet for many, it will be the Victory Air Museum of North America, based at Polidori Airstrip, Mundelein, Illinois, with Paul Polidori as director and Earl Reinert as founder curator, which will be the main attraction if only limited time is available. Here, 20 miles NW of O'Hare Field, Chicago's main airport, in varying stages of restoration, are Republic P-47-M1 Thunderbolt 227385 (NX4477N), Grumman FM2 Wildcat Bu 46867 dedicated to "Butch O'Hare and Jo Foss", Lockheed 18 2064

The Sea Vampire displayed in the Fleet Air Arm Museum at Yeovilton was the first jet aircraft to land on and take off from the deck of a carrier, HMS *Ocean*, on December 3rd 1945

Line-up of some of the full-size aircraft preserved at Yeovilton: left to right, Sea Hawk, Attacker, Sea Fury, Firefly, Corsair, Seafire, Martlet, Swordfish and Walrus. In front of them is a Phantom, latest and last fixed-wing fighter of the FAA

Messerschmitt Me 262B-1A-U1 two-seat night fighter, minus nose antennae, at Willow Grove

CAC CA-1 Wirraway general-purpose aircraft, developed from the Harvard; Moorabbin, Melbourne

Hanriot HD-1 fighter (120hp Le Rhone 9 Jb); in storage awaiting transfer to Lucerne

(CF-TCY), now displayed as a Hudson of RAF Coastal Command, marked ER for Queen Elizabeth II and V for Churchill and dedicated to the Commonwealth Air Forces of World War II, Douglas RB-26C-40-DT Invader 44-355090, Grumman F6F-3 Hellcat Bu 40467 (ex-carrier *Lexington*), North American P-51K/F-6 Mustang 44-12840 (fuselage only), Consolidated PBY-6A Bu 64002 N331RS (being restored), Ohka II suicide aircraft (one of six captured at Okinawa, 1945), North American T-28A 49-1738 (N7692C), Grumman F8F-2 Bearcat Bu 122629, Beech AT-11 N81Y, Chance Vought F8U Cutlass (incomplete), a North American F-86F Sabre fuselage, a Lockheed P-38J (incomplete), Consolidated SNV-1 Bu 34491, Lockheed T-33A-510 51-4298 (fuselage), a Brewster Buffalo and North American SNJ-6 N9826C. There are also the remains of an Me 262, Me 163, Hs 129 and V-1 flying bomb engine.

Over the North Pole now, to call in at Monino, 30 miles from Moscow, to see the Air Forces' Red Banner Academy where, in the grounds, are an Antonov An-2 twelve-seat biplane of 1947, an Ilyushin Il-10 and Ilyushin Il-28 "04", Lavochkin La-7 No 27, credited with 63 "kills" by Ivan Kozedub, Lavochkin La-11 No 20, the last piston-engined fighter designed by Sergei Lavochkin, Lavochkin La-15, powered by the Russian version of the Rolls-Royce Derwent, Mikoyan MiG-9 "01" jet fighter, Mikoyan MiG-15 No 27 (with another in cutaway), Mikoyan MiG-17, Petlyakov Pe-2 of World War II, Polikarpov Po-2, Tupolev Tu-2, Yakovlev Yak-12 communications aircraft, Yakovlev Yak-17 and Yak-23 fighters, and Yakovlev Yak-24 tandem-rotor helicopter. Under cover are a Voisin pusher of World War I and a Sopwith triplane.

In Moscow itself, the Central House of Aviation and Cosmonautics, in addition to some excellent models, two thousand pictures and aviation journals *dating back to* 1881, has a MiG-15 fighter, Mil Mi-1 helicopter and Ka-18 helicopter CCCP 31300; whilst the Central Museum of the Armed Forces displays the remains of a Heinkel He 111, the wreckage of Gary Powers' much-publicised U-2 "spy-plane" and a MiG-15 and MiG-17, whilst, outside, is a MiG-21F serialled "01".

It is, therefore, well worth the stopover in the Soviet Union before we touch down in London, or at whichever airport we boarded for our round-the-world visit to some of the lesser-known, but growing aircraft collections.

Greenland's Ice-Reconnaissance Flights

by Carl Christian Brunckhorst

CLIMATICALLY, the seas around Greenland (Denmark's northernmost province and the world's largest island, of 840,000 square miles) have a mixed reputation: calm and fog change with almost lightning rapidity to hurricane-strength blizzards. The ever-present danger of ice is another complication.

The ice situation around the island, which forms the greatest threat to shipping, has four basic components: ice masses from the Polar Basin (chiefly floes, packed and loose ice up to 10ft thick); one-year old winter-ice of coastal origin, seldom more than 27 inches thick; icebergs detached from the 2-3 miles-deep inland icecap's innumerable glaciers which terminate in the sea (85% of the island is ice-covered); and the ocean and coastal currents which, by and large, move southward along the east coast, around the

southernmost point of Cape Farewell and then north again.

For many years, ships officers navigating these waters were forced to rely on skill acquired through hard experience and on considerable luck. But the loss off Cape Farewell on January 30th, 1959, of the Royal Greenland Trading Department (RGTD) motor vessel *Hans Hedtoft*, with all 95 persons on board, brought drastic changes.

The reason for the loss of the *Hans Hedtoft* on its maiden voyage—a vessel with the latest innovations and specially built for the route—was never established. SOS signals and garbled radio messages were received stating that the ship was sinking rapidly in a severe storm; ice was mentioned, but no details got through. It was suspected that, in spite of the vessel's heavy hull-reinforcement, pressure ridges in the surrounding

Kutdlek Loran station is located on a level spot about half-way between the sea and the peak above the promontary in the picture's centre. It is supplied by air-drop, and any packages that miss the drop-zone can be retrieved only by a long climb

Glacier outlet to the sea on Upernavik island

Remains of US floating ice observation station ARLISS

polar ice, created by hurricanes, crushed the sides of the ship. Another theory is that high winds forced the vessel backward until it collided stern-first, where the hull-reinforcement was weakest, with heavy ice. Rescue attempts were impossible because of the conditions.

Because of this catastrophe and the increasing sea-traffic to and from Greenland, the Danish government, on the recommendations of a commission headed by Vice-Admiral A. H. Vedel, RDN, decided to establish a year-round air reconnaissance system linked with the existing radio information service.

The new service did not have to start from scratch. From its founding in 1872, Denmark's Meteorological Institute had collected data concerning Greenland's ice and had become a world authority on arctic meteorological conditions. Captains of vessels bound for the island received (and still receive) blank charts on which to record ice conditions encountered. Telegraph and other stations along the shores keep logs of the ice. Since 1890 the Institute has published a yearly Greenland ice-report resumé, and information on ice conditions in the Polar Basin, based on observations by nearly all

vessels sailing in those waters, were compiled and published yearly from 1900 to 1956, except for the war years. Since World War II, increased traffic, more coastal stations and, particularly, data from USAF ice-reconnaissance flights have all added to the practical and scientific knowledge of the arctic ice system.

In 1948 the Meteorological Institute had proposed the inauguration of a Greenland radio ice-reporting service, to broadcast from Angmagsalik. Operational from February 15th, 1950, the service had to rely mainly on reports from coastal stations, which always suffered from limited visibility, and on the relatively few ship and aircraft sightings. For various reasons USAF data proved difficult to utilise, and RDAF ice-reconnaissance sorties were made only off Mestersvig on the east coast during the short summer.

Mestersvig was then a highly profitable lead mine, worked on a year-round basis, but with a shipping season of only 3-3½ months. The RDAF Catalinas used for reconnaissance proved invaluable in maintaining the tight schedules of the J. Lauritzen Company's ice-going cargo vessels. In general, however, the radio service was inadequate, but it remained the only aid until the ice-reconnaissance flights directed by the new "Ice Central" of the Meteorological Institute went into operation.

Base for the new organization was Narssarssuaq in SW Greenland, where US forces had constructed an airfield code-named Bluie West I, with a single 7,250ft concrete runway, during World War II. In the autumn of 1959 the RDAF stationed two Catalinas at Narssarssuaq and the first ice-reconnaissance was flown on November 30 of that year. Difficulties in obtaining a sufficient number of experienced aircrews led to the charter of a Catalina operated by Eastern Provincial Airways of Gander during December 1960 and January 1961, after which the aircraft was replaced by the Icelandair DC-4 *Solfaxi*. This arrangement was technically an improvement because, although Catalinas were well suited to the task due to their powerful radio-equipment and excellent visibility from the two side blisters, the DC-4's spacious cabin and four engines were assets appreciated by airmen flying in such a region. Also, the higher take-off weight enabled take-offs and landings to be made under more difficult meteorological conditions.

Solfaxi continued doing fine work on this exacting job until the night of October 23/24th, 1963, when the hangar at Narssarssuaq (for unknown reasons) was completely destroyed by fire, together with the DC-4, two RDAF Catalinas and one American lightplane. In 1967, the contract with Icelandair terminated and the ice flights have since been flown with a DC-4 chartered from Bergen Air Transport. The aircraft used is, incidentally, the late General Eisenhower's wartime *Columbine*.

The standard reconnaissance route is from Narssarssuaq to Frederikshåb on the west coast and then via Cape Farewell to Tingmiarmiut in the east. It is flown with the primary object of establishing the location, direction of movement and composition of drifting polar ice. When the ice extends north of Frederikshåb, the flights are continued until the additional area has been covered.

Once or twice each month flights along the east coast reach Angmagsalik. The normal frequency is two flights per week, the operation being increased during severe conditions and reduced to one a week in the autumn, when the polar ice usually disappears completely. An investigation of the so-called West Ice, originating off Baffin and Ellesmere Islands and then drifting southward via central Baffin Bay and the Davis Straits, was begun a few years ago with flights out of Sondrestrom to Umanak and occasionally Upernavik.

Although the Mestersvig lead mine is now exhausted, ice-reconnaissance flights controlled by Ice Central Mestersvig are still made from July to October by an RDAF Catalina stationed at the field (which has one 6,500-ft gravel runway), chiefly to assist sea-traffic to and from the meteorological stations at Daneborg and Danmarkshavn. These flights are backed up at times by the Narssarssuaq DC-4, which occasionally continues its flight as far as Station Nord, the world's most northerly permanently-manned meteorological station, for combined reconnaissance and supply purposes.

The first Danish ice-reconnaissance flight covering the entire circumference of Greenland was made in May 1968. The flight began on the 24th, with take-off from Narssarssuaq at 12.23 hours GMT. Flying via Cape Farewell, the DC-4 arrived at Reykjavik at 18.13. At 10.35 the following morning it left with members of the 5th Peary Land Expedition, and arrived at Station Nord at 17.36 hours.

Take-off was at 20.15 on the same day, the flight continuing to the Peary Land Expedition HQ at Brønlunds Fjord, where ten 55-lb sacks of coal were dropped from the

Narssarssuaq tower and the remains of the hangar destroyed by fire in 1963. In foreground stands a DC-6B of Icelandair

air. The 'plane then continued to the areas south and west of Brønlundsfjord, where three supply caches were parachuted along the Expedition's later routes. After dropping a further ten sacks of coal at the HQ, the DC-4 landed at 22.10 on an unprepared site near Cape Harald Moltke—an old dried-out fjord of smooth clay. On site to greet the crew was the leader of the expedition, Count Eigil Knuth, who together with two companions had trekked the 175 miles from Station Nord on a snow scooter which was then used to mark the direction in which the aircraft should land. Take-off from Cape Moltke was at 23.04 hours, with arrival back at Station Nord at 23.47.

On the following day, May 25th, the DC-4 departed at 11.55 and, flying via Greenland's north and west coasts, arrived at Thule at 17.10. It left on May 26th at 13.17, and arrived at Sondrestrom at 19.05. After departure at 12.41 hours on May 27th, the DC-4 finally touched down at Narssarssuaq at 19.52 hours.

Ice-reconnaissance DC-4's carry a crew of five or six, plus an ice-observer—a specially trained RGTD ship's officer with considerable experience in Greenland navigation and knowledge of ship construction. The senior observer is also chief of the Ice Central. The normal tour of duty is 24 to 30 months, but the chief is under a longer contract. Reconnaissance flight routings and times are planned jointly by the Ice Central and the pilots concerned, and are based on data from previous flights, the latest land and sea reports, the positions of incoming and outgoing shipping, and on the prevailing meteorological conditions in the area.

En route ice data are plotted by the observer on 1 : 1,000,000 Lambert's Conical Projection maps, based on US Navy Hydrographic Office Ice Plotting Sheets. As a result of the navigator's position fixes and

A Greenlandair S-61N helicopter, used for passenger and supply services to civilian centres and military radar sites

Open water at last—an outward-bound freighter (centre left) noses through the last ice off Cape Farewell

the observer's visual and radar sightings, a mapped picture of ice conditions along the plane's track gradually emerges. Observations are however dependent on visibility: in fog or during low ceiling conditions only radar can be used and this defines only an ice-area's extent and not its composition.

Ice reports based on land, sea and air observation are broadcast on request in plain English via Angmagsalik Radio. Additionally, air-based ice reports are telegraphed from Narssarssuaq to Prins Christiansund, Julianehåb, Grønnedal (the RDN Greenland HQ) and Frederikshåb, from where the reports are re-broadcast on request. The messages include date of the reconnaissance flight, route or observation area, position of the Polar and/or west ice, ice density in tenths, leads, shore-leads, winter-ice and the estimated number of bergs. A radio service based on flights covering the Cape Farewell area is sent twice daily from Narssarssuaq Radio, but is not too successful, due to technical problems.

The ice-reconnaissance 'planes are often used for piloting vessels through ice. On request, an aircraft will meet a ship at a pre-arranged position off the edge of an ice-belt and then guide the vessel through the most suitable leads and stretches of open water. Shore-based helicopters are also used for this work. During recent years, ships sailing to outposts in Greenland on regular schedules have found it a paying proposition to carry a chartered helicopter, usually on a specially-built deck platform. Normally the helicopter can guide the parent vessel into harbour in a matter of hours under conditions in which ships without aid might spend days trying to penetrate the ice.

The ice-information organization only recently became a permanent establishment, due chiefly to uncertainty regarding the future of Narssarssuaq airfield. After the hangar fire, a temporary wall of locally-available material, affording some protection, was constructed, but the lack of hangarage often causes take-off delays due to lengthy de-icing and snow removal operations. Reports via Narssarssuaq Radio at times contain errors and the Ice Central maintains rightly that a more powerful transmitter is needed before this base can offer a first-class service capable of reaching a large number of ships. The main problem is financial. Nearly all building material and equipment has to be transported from Europe or the USA and erected or installed by personnel from overseas, drawing high salaries. Despite all the problems, the present service is a great improvement and is much appreciated by those who travel on the waters around Greenland.

Bergen Air Transport's ice-reconnaissance DC-4, named *Moby Dick*, at Narssarssuaq

Journey Round the Moon *by Maurice Allward*

Apollo 8—Interplanetary Venture

WE, THE PEOPLE of our time, were privileged in December 1968 to witness one of the great moments in the history of mankind when Apollo 8 circled the Moon with astronauts Frank Borman, James Lovell and William Anders on board. There have been previous long and exciting voyages of discovery but none was so long or so momentous as the flight of Apollo 8. It was Man's first-ever interplanetary venture.

The fact that the voyage was almost completely free from snags or moments of great anxiety tended to belie the magnitude of this, the greatest technical achievement of all time, requiring the combined resources of all the sciences.

The countdown began several days before the launch. This revealed a number of faults, all of which were rectified without loss of time, as the countdown programme had three planned 'holds' built in for just such eventualities.

On the day of the launch, Saturday, December 21st, the three astronauts, after a final physical check-up and breakfast, entered the command module at 10.34 hours, and the final countdown started. The launch was planned for 13.51 and the remaining

---

View from Apollo 8 as it approached the Moon.   The circular dark-coloured area at the upper centre is the Sea of Crises.   Most of the right half of this picture shows the side of the Moon which is never visible from Earth

checks continued uneventfully. The final 20 minutes of the countdown were televised, to what must surely have been the largest audience of all time. There is no doubt that this 'instant TV' added to the strain and problems associated with the flight. Full credit, then, is due to the officials who made the decision that enabled millions of ordinary people all over the world to share with America the excitement of her great venture. To help disseminate the story, no fewer than 1,550 press men were accommodated at NASA's Manned Spacecraft Center, at Houston, Texas.

At T minus 6 seconds, the five F-1 engines of the first stage ignited and built up to full thrust. Then, at 13.51, Apollo 8 started to lift from the pad—one sixth of a second late, apparently!

After two minutes, the centre F-1 engine was shut down to prevent the acceleration from exceeding 4g. Separation of the first S.IC stage took place at an altitude of about 50 miles, 54 miles down-range at a speed of 6,068 mph, and was clearly visible from the ground, together with the ignition of the five J-2 engines in the S.II second stage. To TV watchers the whole craft was momentarily enveloped in a heart-stopping cloud of smoke and flame. Later, when seeing a telerecording of the event, Frank Borman, commander of the mission, remarked that he was glad he wasn't able to see it at the time!

Two orbits of the Earth were made, during which the complete spacecraft was checked out. The Manned Spacecraft Center then instructed: "Apollo 8, you are to go for TLI" (trans-lunar injection). The S.IVB stage was fired for 5 minutes to accelerate the spacecraft to 24,227 mph—"exactly nominal" according to delighted officials at Houston.

From this moment, the flight of Apollo 8 was subtly different from that of any previous mission, either American or Russian. Previously, in the event of trouble, astronauts could initiate emergency re-entry procedures at any time and were only a few hours away from help. Apollo 8, on the other hand, was on a ballistic trajectory which would carry it round the Moon before it could return to Earth. The crew were thus very much on their own, and this had a significant effect on their attitude.

In the early hours of Sunday morning (British time) Borman fired the 20,500lb thrust engine in the service module for the first of four possible mid-course correction manoeuvres, to increase the speed of the spacecraft by 17mph.

Later on Sunday all three astronauts felt sick and had stomach pains, Borman being particularly upset, and were prescribed pills by their doctor at Houston. This illness was subsequently the subject of outspoken comment by doctors associated with the Apollo programme, when colds delayed the launch of Apollo 9 in February 1969. It was, they claim, the inevitable consequence of the intensive and prolonged training programme and over-work during space missions.

In spite of the illness, which must have been more worrying than the astronauts like to admit, the first of several live TV broadcasts was made on Sunday evening, when the spacecraft was about 135,000 miles from Earth. The transmission was made using a portable RCA camera which, when held against a window, showed viewers in their homes the astronaut's view of the Earth from space. These initial pictures were rather blurred, but the clarity of speech was outstanding. During this transmission the astronauts commented that some of the windows had misted up, and at one time during the flight three of the five windows became obscured. This window fogging has been a problem since the Gemini days, and it is rather surprising that it reappeared on Apollo, as clear vision is vital for navigation fixes.

Public interest increased sharply after this

Apollo 8, carrying astronauts Frank Borman, James Lovell and William Anders, gets away to a spectacular start on Saturday, December 21st 1968

Frank Borman waves to viewers at the end of the first live TV transmission from Apollo 8 on December 22nd

During the first transmission, James Lovell wishes his mother "Happy Birthday", 135,000 miles from Earth

William Anders uses his toothbrush to demonstrate weightlessness. Apollo 8's speed at this stage was 3,207 mph

TV broadcast, and the progress of the flight cropped up more and more in everyday conversation. People began to feel that, because they had seen the astronauts at work and play, and because of the views of the Earth from space, they were taking part in the mission personally.

On Monday evening another transmission was made, showing the Moon from a distance of 100 miles. To avoid vibration the camera was held against the structure, subsequent aiming being achieved by orientating the entire spacecraft. The quality of this broadcast was outstanding, and viewers on Earth were able to see the effect of different filters on the camera as these were changed on commands from Mission Control.

At this point one of the several moments of high tension was near—the decision to attempt to go into lunar orbit or simply to let the craft swing round the Moon and head back to Earth. The condition of the spacecraft was satisfactory and Houston radioed the historic message: "This is Houston at 68.04 (68 hours 4 minutes after launch); you are go for LOI" (Lunar orbit insertion). Borman replied "OK, Apollo 8 is go". The service module engine was fired for just over 4 minutes to decrease the speed from 5,700 mph to 3,600 mph and so place the craft in an initial elliptical lunar parking orbit.

This critical burn took place behind the Moon, when the craft was out of radio contact with the Earth, and tension rose everywhere. Borman's heart-beats rose from 78 per minute to 130 per minute. That of the author, in the safety of his home, was I am sure even higher! The craft was out of contact for about half an hour, and for this period the anxious controllers at Houston did not know if the engine had performed satisfactorily. If not, the craft could either have gone into eternal orbit round the Sun, or have crashed into the rear of the Moon.

Eventually, to everyone's relief, Apollo 8 appeared over the lunar horizon. The actual moment seems to have been marked by the technical request from Houston: "Please verify your water-evaporation switch to auto . . . ." To the waiting world Houston reported: "We've acquired a signal but no voice yet. We are looking at engine data and it looks good. Tank pressures look good. We got it! Apollo 8 is in lunar orbit."

After their first orbit of the Moon, the astronauts were greeted by this majestic view of the Earth, with the sunset terminator bisecting Africa. About 100 miles of Moon horizon are visible

From the spacecraft, astronaut Lovell replied: "Good to hear your voice".

Understandable jubilation, including cheering, filled the Houston control room. Flight Director Glynn Lunney commented afterwards: "It certainly wasn't a faint reaction. There was quite a bit of racket. I'm sure it can be described as one of the happiest Christmas Eves anyone there had seen".

After two orbits, the service module engine was fired again to place the craft in a 69-mile circular orbit, a manoeuvre requiring precise orientation and a speed reduction of only 138 ft/sec.

Thus, Man was at the Moon, bringing true the dreams of untold generations. From their grandstand view, telecasts gave Earth-bound viewers an unforgettable astronaut's-eye view, and the first ever close-up description of the surface.

"The Moon is essentially grey, no colour" reported Lovell. "Looks like plaster of Paris, or sort of a greyish beach sand . . . "

" . . . The Sea of Fertility doesn't stand out as well here as it does back on Earth . . .

"The craters are all rounded off. There's quite a few of them. Some are newer. Many of them look like—especially the round ones—look like they have been hit by meteorites or projectiles of some sort . . . Langrenus is quite a huge crater. Its got a central core to it. The walls of the crater are terraced, about six or seven terraces on the way down."

While in orbit the primary task of the flight was undertaken—photography of the Moon and particularly of the sites which have been selected for the first landing. Some of the photographs obtained are in colour and are expected to be of value in determining the nature and topography of the surface. Although the surface is essentially grey, the photographs show a distinct green shading; this is thought to be either due to the film or to the inter-reaction of the cabin windows and the camera lens.

In reporting their observations the astronauts referred to a number of craters by such names as Mercury, Apollo, Slayton, Grissom, White and Chaffee. Houston officials explained that the names were a purely unofficial code to help identify new craters.

Oblique photograph of the hidden side of the Moon taken from Apollo 8. The horizon is about 275 miles away

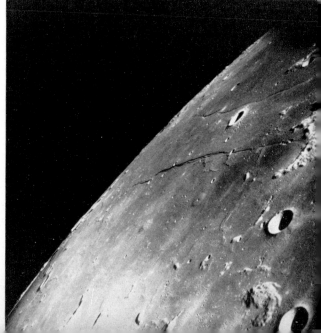

Part of the Sea of Tranquility. Of interest are two parallel rills, or faults, with the Cauchy crater between them

However, three craters in a group were, appropriately, named Borman, Lovell and Anders, and it is likely that these names will be officially adopted, as well as that of a mountain christened Marilyn by Lovell, after his wife. After the flight Borman said he was sorry that he did not have the foresight to name a crater in honour of Komarov, the Russian cosmonaut who died while testing the new Soyuz spacecraft.

Frank Borman's impression of the Moon was "a vast forbidding type of expanse of nothing . . . not a very exciting place to live or work".

James Lovell thought the Moon was "awe-inspiring . . . It makes you realize just what you have back there on Earth. The Earth from here is a grand oasis in the big vastness of space." Lovell also reported a lunar phenomenon that puzzles officials. "Before the Sun came above the limb (horizon) definite rays could be seen coming from it. It was a uniform haze apparently where the Sun was going to rise." The observation suggests that the Moon might have a tenuous atmosphere, contrary to previous thought, and this undoubtedly will be investigated on future flights.

A second TV broadcast was made during the ninth orbit, in the evening. As the camera passed over the desolate scene, William Anders said "We are now approaching the lunar sunrise and for all the people back on Earth the crew of Apollo 8 have a message that we would like to send you".

And then, in turn, the three astronauts read the first 10 verses of Chapter 1 from the Book of Genesis, the passage in the Bible describing the Creation of the World. For people of Jewish and Christian Faith the message was particularly appropriate but, read with becoming humility, it also captured the significance of the occasion for all listeners.

The last of the ten planned orbits was spent largely in preparing for the trans-Earth injection manoeuvre (TEI) which would accelerate the craft to 5,980mph to start it on the long journey home.

To many this was the most agonising moment of all. If the service module engine had not fired on the approach to the Moon, the craft would merely have looped round it and returned to Earth. Now, in lunar orbit, if it failed to start Apollo 8 would have been doomed to circle the Moon for ever, its crew dying when their oxygen gave out.

There were no jokes as the craft approached the edge of the Moon. Houston indicated that everything seemed all right, with the message: "All systems are go, Apollo 8". "Roger" Frank Borman acknowledged tersely.

The critical burn was initiated behind the Moon. The motor fired and burned for 3 minutes 46 seconds, as planned. Houston—and the world—had to wait an agonising 37 minutes until the craft came into view, heading for home.

As the craft emerged, astronaut Lovell reported: "Please be informed that there is a Santa Claus."

The orientation and length of burn were so accurate that only one of three midcourse correction manoeuvres was necessary. On the return journey two more live TV broadcasts were made and, to reduce the monotony during off-duty periods, the astronauts were entertained with music, football scores and lengthy newscasts. They accepted, with good nature, the claim by London's Flat Earth Society that the public was being ballyhooed, and taken for a ride. Borman retorted: "It doesn't look flat from here, but, I don't know—maybe something is wrong with our vision".

During the final TV broadcast on Boxing Day, Borman said the astronauts felt something like the travellers of old after a very long voyage away from home. "Now we are heading back we feel proud of our flight—but we're glad to be heading back for home".

The astronauts ended with: "We will see

A MESSAGE FOR CHRISTMAS

One of the surprises of the flight of Apollo 8 was a broadcast on Christmas Eve of a reading from the Book of Genesis, the passage in the Bible describing the Creation of the World.

From Major Anders came the words:

"In the beginning, God created the heaven and the earth. And the earth was without form, and void; and darkness was upon the face of the deep. And the Spirit of God moved upon the face of the waters. And God said, Let there be light: and there was light. And God saw the light, that it was good: and God divided the light from the darkness."

Then, from 230,000 miles out in space, the voice of Captain Lovell continued:

"And God called the light Day, and the darkness He called Night. And the evening and the morning were the first day. And God said, Let there be a firmament in the midst of the waters, and let it divide the waters from the waters.

"And God made the firmament, and divided the waters which were under the firmament from the waters which were above the firmament: and it was so. And God called the firmament Heaven and the evening and the morning were the second day."

Colonel Borman completed the reading with the verses:

"And God said. Let the waters under the heaven be gathered together unto one place, and let the dry land appear: and it was so. And God called the dry land Earth; and the gathering together of the waters called He Seas: and God saw that it was good . . . "

Colonel Borman continued: *"And from the crew of Apollo 8, we close with good night, good luck, a merry Christmas, and God bless all of you—all of you on the good Earth."*

The decision to insert the Bible reading into the Apollo flight plan was made at quite a late stage, at a pre-launch luncheon at Cape Kennedy. The relevant verses were typed on fireproof paper and read out; a non-fireproof Bible was carried on board in a fireproof container, but was difficult to get at in flight. After the flight Colonel Borman explained he had been told that he would have the largest audience of all time and that he was to put it to good use. It is difficult to think of a more appropriate message which, coming with remarkable clarity from the vicinity of the Moon, was for many the most moving moment of the whole historic flight.

---

you back on the good Earth very soon."

Nearing home, Apollo 8 began to accelerate to a maximum speed of 24,530mph. During the last few hours the crew stowed away all loose gear. One of the final acts was to jettison the service module—with its faultless SPS engine—leaving only the command module, which was then rotated so that its blunt ablative base faced the direction of flight.

The danger and tension were not yet over.

One major hurdle remained: the re-entry into the atmosphere at a speed some 7,000 mph faster than the re-entry speed of any previous manned spacecraft. Apollo 8 had to strike the atmosphere at an angle no greater than 7·4 degrees nor less than 5·4 degrees. If it re-entered too steeply, excessive deceleration forces would probably cause the craft to break up. In a too-shallow re-entry the craft would probably bounce off the atmosphere back into space, where the

Safely in their rubber dinghy, the three astronauts wait for a helicopter to take them to the prime recovery vessel, the carrier USS *Yorktown*. The spacecraft ditched only 2-3 miles from the ship, within 11 seconds of the time predicted at the time of the launch, six days previously

spacecraft's fiery track, five miles wide and an incredible 100 miles long. After penetrating to 200,000ft, the craft roller-coasted up several thousand feet to allow some of the kinetic heat to dissipate. The command module then began its final descent, the speed continuing to fall until, at 23,000ft, when it was only 300mph, the pilot parachute was deployed. The three main parachutes opened at 10,000ft and the craft splashed down in the sea at 16·51 (British time) within two or three miles of the primary recovery ship, the aircraft carrier USS *Yorktown*. The landing was within 11 seconds of that predicted, and exactly 147 hours after the spectacular launch.

The splashdown occurred in darkness and the crew elected to stay in the craft until daylight, when they were hoisted, one at a time, aboard a recovery helicopter and flown to the carrier.

On board, Borman smiled and waved to acknowledge the cheers from hundreds of sailors—and the thanks-for-a-safe-return thoughts of hundreds of millions of TV viewers.

The world acknowledged the bravery, daring and modesty of the astronauts, President Johnson telephoned: "You have made us feel kin to those Europeans five centuries ago who first heard news of the New World. You've seen what man has never seen before".

oxygen supplies would probably be exhausted before the craft re-entered again and landed.

As it was, the craft entered at almost precisely the planned angle of 6·43 degrees.

As it sped through the atmosphere the temperature of the heat shield rose to a searing 5,000 deg.F. A pilot, flying over the Pacific, reported that he could see the

In the years to come there will, of course, be other exciting voyages through space. In fact, by the time this article appears in print astronauts may have landed on the Moon. There will also be longer voyages, to the planets and, perhaps, even to the nearer stars. But none can ever share what is Apollo 8's special claim to fame. It took mankind on its first-ever interplanetary flight.